I0486459

# The Sega Saturn

A Comprehensive Look at the History and Technology of the Saturn

# Contents

# Chapter 1

# Introduction

## 1.1 Sega Saturn

The **Sega Saturn** (セガサターン *Sega Satān*) is a 32-bit fifth-generation home video game console that was developed by Sega and released on November 22, 1994 in Japan, May 11, 1995 in North America, and July 8, 1995 in Europe as the successor to the successful Sega Genesis. Saturn has a dual-CPU architecture and a total of eight processors. Its games are in CD-ROM format, and its game library contains several arcade ports as well as original titles.

Development of the Saturn began in 1992, the same year Sega's groundbreaking 3D Model 1 arcade hardware debuted. Designed around a new CPU from Japanese electronics company Hitachi, another video display processor was incorporated into the system's design in early 1994 to better compete with Sony's forthcoming PlayStation. The Saturn was initially successful in Japan, but failed to sell in large numbers in the United States after its surprise May 1995 launch, four months before its scheduled release date. After the debut of the Nintendo 64 in late 1996, the Saturn rapidly lost market share in the U.S., where it was discontinued in 1999. Having sold 9.5 million units worldwide, the Saturn is considered a commercial failure. The failure of Sega's development teams to release a game in the *Sonic the Hedgehog* series, known in development as *Sonic X-treme*, has been considered a factor in the console's poor performance.

Although the system is remembered for several well-regarded games, including *Nights into Dreams...*, the *Panzer Dragoon* series, and the *Virtua Fighter* series, the Saturn's reception is mixed due to its complex hardware design and limited third-party support. Sega's management has been criticized for its decision-making during the system's development and cancellation.

### 1.1.1  History

**Background**

Released in 1988, the Sega Genesis (known as the Sega Mega Drive in Europe and Japan) was Sega's entry into the fourth generation of video game consoles.[*][2] In mid-1990, Sega CEO Hayao Nakayama hired Tom Kalinske as President and CEO of Sega of America. Kalinske developed a four-point plan for sales of the Genesis: lower the price of the console, create a U.S.-based team to develop games targeted at the American market, continue aggressive advertising campaigns, and sell *Sonic the Hedgehog* with the console.[*][3] The Japanese board of directors initially disapproved of the plan,[*][4] but all four points were approved by Nakayama, who told Kalinske, "I hired you to make the decisions for Europe and the Americas, so go ahead and do it." [*][2] Magazines praised *Sonic* as one of the greatest games yet made, and Sega's console finally took off as customers who had been waiting for the Super Nintendo Entertainment System (SNES) decided to purchase a Genesis instead.[*][5] However, the release of a CD-based add-on for the Genesis, the Sega CD (known as Mega-CD outside of North America), had been commercially disappointing.[*][6][*][7]

Sega also experienced success with arcade games. In 1992 and 1993, the company's new Sega Model 1 arcade system board showcased Sega AM2's *Virtua Racing* and *Virtua Fighter* (the first 3D fighting game), which played a crucial role in popularizing 3D polygonal graphics.[*][8][*][9][*][10] In particular, *Virtua Fighter* garnered praise for its simple three-button control scheme, with the game's strategy coming from the intuitively observed differences between characters that felt and acted differently rather than the more ornate combos of 2D competitors. Despite its crude visuals—with characters composed of fewer than 1,200 polygons—*Virtua Fighter*'s fluid animation and relatively realistic depiction of distinct fighting styles gave its combatants a lifelike presence considered impossible to replicate with sprites.[*][11][*][12][*][13] The Model 1 was an expensive system board, and bringing home releases of its games to the Genesis required more than its hardware could handle. Several alternatives helped to bring Sega's newest arcade games to the console, such as the Sega Virtua Processor chip used for *Virtua Racing*, and eventually the Sega 32X add-on.[*][14]

## Development

Development of the Saturn was supervised by Hideki Sato, Sega's director and deputy general manager of research and development.[*][15] According to Sega project manager Hideki Okamura, the Saturn project started over two years before the system was showcased at the Tokyo Toy Show in June 1994. The name "Saturn" was initially the system's codename during development in Japan, but was eventually chosen as the official product name.[*][16] In 1993, Sega and Japanese electronics company Hitachi formed a joint venture to develop a new CPU for the Saturn, which resulted in the creation of the "SuperH RISC Engine" (or SH-2) later that year.[*][17][*][18] The Saturn was ultimately designed around a dual-SH2 configuration. According to Kazuhiro Hamada, Sega's section chief for Saturn development during the system's conception, "the SH-2 was chosen for reasons of cost and efficiency. The chip has a calculation system similar to a DSP [digital signal processor], but we realized that a single CPU would not be enough to calculate a 3D world." [*][17][*][19] Although the Saturn's design was largely finished before the end of 1993, reports in early 1994 of the technical capabilities of Sony's upcoming PlayStation console prompted Sega to include another video display processor (VDP) to improve the system's 2D performance and texture-mapping.[*][17][*][19][*][20] CD-ROM-based and cartridge-only versions of the Saturn hardware were considered for simultaneous release at one point during the system's development, but this idea was discarded due to concerns over the lower quality and higher price of cartridge-based games.[*][17]

According to Kalinske, Sega of America "fought against the architecture of Saturn for quite some time".[*][21] Seeking an alternative graphics chip for the Saturn, Kalinske attempted to broker a deal with Silicon Graphics, but Sega of Japan rejected the proposal.[*][22][*][23][*][24] Silicon Graphics subsequently collaborated with Nintendo on the Nintendo 64.[*][22][*][25] Kalinske, Sony Electronic Publishing's Olaf Olafsson, and Sony America's Micky Schulhof had previously discussed development of a joint "Sega/Sony hardware system", which never came to fruition due to Sega's desire to create hardware that could accommodate both 2D and 3D visuals and Sony's competing notion of focusing entirely on 3D technology.[*][23][*][26][*][27] Publicly, Kalinske defended the Saturn's design: "Our people feel that they need the multiprocessing to be able to bring to the home what we're doing next year in the arcades." [*][28]

In 1993, Sega restructured its internal studios in preparation for the Saturn's launch. To ensure high-quality 3D games would be available early in the Saturn's life, and to create a more energetic working environment, developers from Sega's arcade division were instructed to create console games. New teams, such as *Panzer Dragoon* developer Team Andromeda, were formed during this time.[*][29]

In January 1994, Sega began to develop an add-on for the Genesis, the Sega 32X, which would serve as a less-expensive entry into the 32-bit era. The decision to create the add-on was made by Nakayama and widely supported by Sega of America employees.[*][6] According to former Sega of America producer Scot Bayless, Nakayama was worried that the Saturn would not be available until after 1994 and that the recently released Atari Jaguar would reduce Sega's hardware sales. As a result, Nakayama ordered his engineers to have the system ready for launch by the end of the year.[*][6] The 32X would not be compatible with the Saturn, but Sega executive Richard Brudvik-Lindner pointed out that the 32X would play Genesis titles, and had the same system architecture as the Saturn.[*][30] This was justified by Sega's statement that both platforms would run at the same time, and that the 32X would be aimed at players who could not afford the more expensive Saturn.[*][6][*][31] According to Sega of America research and development head Joe Miller, the 32X served a role in assisting development teams to familiarize themselves with the dual SH-2 architecture also used in the Saturn.[*][32] Because both machines shared many of the same parts and were preparing to launch around the same time, tensions emerged between Sega of America and Sega of Japan when the Saturn was given priority.[*][6]

## Launch

*A first model Japanese Sega Saturn unit*

Sega released the Saturn in Japan on November 22, 1994, at a price of JP¥44,800.[*][33] *Virtua Fighter*, a nearly indistinguishable port of the popular arcade game, sold at a nearly one-to-one ratio with the Saturn hardware at launch and was crucial to the system's early success in Japan.[12][13][34] Along with *Virtua Fighter*, Sega had wanted the launch to include both *Clockwork Knight* and *Panzer Dragoon*, but the latter was not ready in time.[29] Aside from *Virtua Fighter*, the only first-party title available on launch day was *Wan Chai Connection*.[35] Fueled by the popularity of *Virtua Fighter*, Sega's initial shipment of 200,000 Saturn units sold out on the first day.[34][36][37] Sega waited until the December 3 launch of the PlayStation to ship more units; when both were sold side-by-side, the Saturn proved to be the more popular system.[34][38] Meanwhile, the 32X was released on November 21, 1994 in North America, December 3, 1994 in Japan, and January 1995 in PAL territories, and was sold at less than half of the Saturn's launch price.[39][40] After the holiday season, however, interest in the 32X rapidly declined.[6][31] Between 400,000 and 500,000 Saturn units were sold in Japan within its first month on the market (compared to 300,000 PlayStation units sold within its first 30 days[41]), and sales exceeded 1 million within the following six months.[42][43] There were conflicting reports that the PlayStation enjoyed a higher sell-through rate, and the system gradually began to overtake the Saturn in sales during 1995.[43] Sony attracted many third-party developers to the PlayStation with a liberal $10 licensing fee, excellent development tools, and the introduction of a revolutionary 7- to 10-day order system that allowed publishers to meet demand more efficiently than the 10- to 12-week lead times for cartridges that had previously been standard in the Japanese video game industry.[44][45]

In March 1995, Sega of America CEO Tom Kalinske announced that the Saturn would be released in the U.S. on "Saturnday" (Saturday) September 2, 1995.[46][47] However, Sega of Japan mandated an early launch to give the Saturn an advantage over the PlayStation.[48] Therefore, at the first Electronic Entertainment Expo (E3) in Los Angeles on May 11, 1995, Kalinske gave a keynote presentation for the upcoming Saturn in which he revealed the release price at US$399 (including a bundled copy of *Virtua Fighter*[49]), and described the features of the console. Kalinske also revealed that, due to "high consumer demand",[50] Sega had already shipped 30,000 Saturns to Toys "R" Us, Babbage's, Electronics Boutique, and Software Etc. for immediate release.[46] This announcement upset retailers who were not informed of the surprise release, including Best Buy and Walmart;[23][51][52] KB Toys responded by dropping Sega from its lineup.[46] Sony subsequently unveiled the retail price for the PlayStation: Sony Computer Entertainment America president Steve Race took the stage, said "$299", and then walked away to applause.[23][53][54] The Saturn's release in Europe also came before the previ-

ously announced North American date, on July 8, 1995, at a price of GB£399.99.[14] European retailers and press did not have time to promote the system or its games, leading to poor sales.[55] After its European launch on September 29, by early November 1995 the PlayStation had already outsold the Saturn by a factor of three in the United Kingdom, where it was reported that Sony allocated £20 million to market the system during the holiday season compared to Sega's £4 million.[56][57]

The Saturn's U.S. launch was accompanied by a reported $50 million advertising campaign that included coverage in publications such as *Wired* and *Playboy*.[42][58][59] Because of the early launch, the Saturn had only six games (all published by Sega) available to start as most third-party games were slated to be released around the original launch date.[60] *Virtua Fighter*'s relative lack of popularity in the West, combined with a release schedule of only two games between the surprise launch and September 1995, prevented Sega from capitalizing on the Saturn's early timing.[21][36][61] Within two days of its September 9, 1995 launch in North America, the PlayStation (backed by a large marketing campaign[44][62]) sold more units than the Saturn had in the five months following its surprise launch, with 100,000 units presold in advance and sell-outs reported throughout the U.S.[43][63]

A high-quality port of the Namco arcade game *Ridge Racer* contributed to the PlayStation's early success,[38][64] and garnered favorable comparisons in the media to the Saturn version of Sega's *Daytona USA*, which was considered inferior to its arcade counterpart.[65][66] Namco, a longtime arcade competitor with Sega,[9][67] also unveiled the Namco System 11 arcade board, which was based on raw PlayStation hardware.[68] Although the System 11 was technically inferior to Sega's Model 2 arcade board, its lower price made it an attractive prospect for smaller arcades.[68][69] Following a 1994 acquisition of Sega developers, Namco released *Tekken* for the System 11 and PlayStation. Directed by former *Virtua Fighter* designer Seiichi Ishii, *Tekken* was intended to be a fundamentally similar title, with the addition of detailed textures and twice the frame rate.[70][71][72] *Tekken* surpassed *Virtua Fighter* in popularity due to its superior graphics and nearly arcade-perfect console port, becoming the first million-selling PlayStation title.[69][73][74] On October 2, 1995 Sega announced a Saturn price reduction to $299.[75] Moreover, high-quality Saturn ports of the Sega Model 2 arcade hits *Sega Rally Championship*,[76] *Virtua Cop*,[77] and *Virtua Fighter 2* (running at 60 frames per second at a high resolution)[78][79][80] were available by the end of the year—and were generally regarded as superior to any competitors on the PlayStation.[14][81] Notwithstanding a subsequent increase in Saturn sales during the 1995 holiday season, these games were not enough to reverse the PlayStation's decisive lead.*[81]*[82] By 1996, the PlayStation had a considerably larger library than the Saturn, although Sega hoped to generate increased interest in the Saturn with upcoming exclusives such as *Nights into Dreams....*\*[61] Within its first year, the PlayStation secured over 20% of the entire U.S. video game market.*[59] At the May 1996 E3 show, Sony announced a PlayStation price reduction to $199, and shortly afterwards Sega decided to match this price, even though Saturn hardware was more expensive to manufacture.*[43]*[83]

### Changes at Sega

"I thought the world of [Hayao] Nakayama because of his love of software. We spoke about building a new hardware platform that I would be very, very involved with, shape the direction of this platform, and hire a new team of people and restructure Sega. That, to me, was a great opportunity."

—Bernie Stolar, on his joining Sega of America.*[36]

In spite of the launch of the PlayStation and the Saturn, sales of 16-bit hardware/software continued to account for 64% of the video game market in 1995.*[84]*[85] Sega underestimated the continued popularity of the Genesis, and did not have the inventory to meet demand for the product.*[82]*[84] Sega was able to capture 43% of the dollar share of the U.S. video game market and sell more than 2 million Genesis units in 1995, but Kalinske estimated that "we could have sold another 300,000 Genesis systems in the November/December timeframe." *[82] Nakayama's decision to focus on the Saturn over the Genesis, based on the systems' relative performance in Japan, has been cited as the major contributing factor in this miscalculation.*[86]

Due to long-standing disagreements with Sega of Japan,*[23]*[36] Kalinske lost most of his interest in his work as CEO of Sega of America.*[87] By the spring of 1996, rumors were circulating that Kalinske planned to leave Sega,*[88] and a July 13 article in the press reported speculation that Sega of Japan was planning significant changes to Sega of America's management team.*[89] On July 16, 1996 Sega announced that Shoichiro Irimajiri had been appointed chairman and CEO of Sega of America, while Kalinske would be leaving Sega after September 30 of that year.*[90]*[91] A former Honda executive,*[92]*[93] Irimajiri had been actively involved with Sega of America since joining Sega in 1993.*[90]*[94] Sega also announced that David Rosen and Nakayama had resigned from their positions as chairman and co-chairman of Sega of America, though both men remained with the company.*[90]*[95] Bernie Stolar, a former executive at Sony Computer Entertainment of America,*[89]*[96] was named Sega of America's executive vice president

in charge of product development and third-party relations.*[90]*[91] Stolar, who had arranged a six-month PlayStation exclusivity deal for *Mortal Kombat 3*\*[97] and helped build close relations with Electronic Arts*[36] while at Sony, was perceived as a major asset by Sega officials.*[91] Finally, Sega of America made plans to expand its PC software business.*[90]*[93]

Stolar was not supportive of the Saturn due to his belief that the hardware was poorly designed, and publicly announced at E3 1997 that "The Saturn is not our future." *[36] While Stolar had "no interest in lying to people" about the Saturn's prospects, he continued to emphasize quality games for the system,*[36] and subsequently reflected that "we tried to wind it down as cleanly as we could for the consumer." *[96] At Sony, Stolar opposed the localization of certain Japanese PlayStation titles that he felt would not represent the system well in North America, and he advocated a similar policy for the Saturn during his time at Sega, although he later sought to distance himself from this perception.*[36]*[97]*[98] These changes were accompanied by a softer image that Sega was beginning to portray in its advertising, including removing the "Sega!" scream and holding press events for the education industry.*[61] Marketing for the Saturn in Japan also changed with the introduction of "Segata Sanshiro" (played by Hiroshi Fujioka) as a character in a series of TV advertisements starting in 1997; the character would eventually star in a Saturn video game.*[99]*[100]

Temporarily abandoning arcade development, Sega AM2 head Yu Suzuki began developing several Saturn-exclusive games, including a role-playing game in the *Virtua Fighter* series.*[101] Initially conceived as an obscure prototype called "The Old Man and the Peach Tree" and intended to address the flaws of contemporary Japanese RPGs (such as poor non-player character artificial intelligence routines), *Virtua Fighter RPG* evolved into a planned 11-part, 45-hour "revenge epic in the tradition of Chinese cinema"–which Suzuki hoped would become the Saturn's killer app.*[36]*[102]*[103] The game was eventually released as *Shenmue* for the Saturn's successor, the Dreamcast.*[104]*[105]

### Cancellation of *Sonic X-treme*

Main article: Sonic X-treme

Sega tasked the U.S.-based Sega Technical Institute (STI) with developing what would have been the first fully 3D entry in its popular *Sonic the Hedgehog* series. The game, known as *Sonic X-treme*, was moved to the Saturn after several prototypes were discarded.*[106]*[107]*[108] Featuring a fisheye lens camera system that caused levels to rotate with Sonic's movement, the project was set

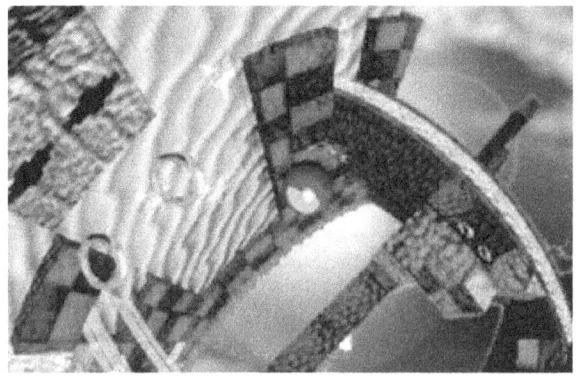

*A screenshot of Chris Senn and Ofer Alon's version of* Sonic X-treme. *The game's cancellation, and the lack of a fully 3D* Sonic the Hedgehog *platformer, is considered a significant factor in the Saturn's commercial failure.*

back after *Sonic* creator Yuji Naka refused to allow the developers access to the engine he created for *Nights into Dreams....*[107][109] Sega of Japan executives who visited STI in March 1996 were unimpressed by *X-treme*'s progress, so Nakayama ordered that the entire game be reworked around the engine created specifically for its boss battles, and employees worked between 16 and 20 hours a day in an attempt to meet their December 1996 deadline.[107][108][109] After programmer Ofer Alon quit and designer Chris Senn caught pneumonia, the project was cancelled in early 1997.[107][108][109] Sonic Team started work on an original 3D *Sonic* title for the Saturn (which eventually became *Sonic Adventure*), but development was shifted to the Dreamcast.[110][111] STI was officially disbanded in 1996 as a result of changes in management at Sega of America.[106]

Journalists and fans have speculated about the impact a completed *X-treme* might have had on the market. David Houghton of GamesRadar described the prospect of "a good 3D *Sonic* game" on the Saturn as "a 'What if...' situation on a par with the dinosaurs not becoming extinct." [108] IGN's Tavis Fahs called *X-treme* "the turning point not only for SEGA's mascot and their 32-bit console, but for the entire company", although he also noted that the game served as "an empty vessel for SEGA's ambitions and the hopes of their fans".[107] Dave Zdyrko, who operated a prominent website for Saturn fans during the system's lifespan, offered a more nuanced perspective: "I don't know if [*X-treme*] could've saved the Saturn, but ... *Sonic* helped make the Genesis and it made absolutely no sense why there wasn't a great new *Sonic* title ready at or near the launch of the [Saturn]".[21] In a 2007 retrospective, producer Mike Wallis maintained that *X-treme* "definitely would have been competitive" with Nintendo's *Super Mario 64*.[109]

## Decline

From 1993 to early 1996, although Sega's revenue declined as part of an industry-wide slowdown,[59][112] the company retained control of 38% of the U.S. video game market (compared to Nintendo's 30% and Sony's 24%).[85] 800,000 PlayStation units were sold in the U.S. by the end of 1995, compared to 400,000 Saturn units.[113][114] In part due to an aggressive price war,[59] the PlayStation outsold the Saturn by two-to-one in 1996, while Sega's 16-bit sales declined markedly.[85] By the end of 1996, the PlayStation had sold 2.9 million units in the U.S., more than twice the 1.2 million units sold by the Saturn.[51] After the launch of the Nintendo 64 in 1996, sales of the Saturn and Sega's 32-bit software were sharply reduced,[96] while the PlayStation outsold the Saturn by three-to-one in the U.S. market in 1997.[59] The 1997 release of *Final Fantasy VII* significantly increased the PlayStation's popularity in Japan.[115][116] As of August 1997, Sony controlled 47% of the console market, Nintendo controlled 40%, and Sega controlled only 12%. Neither price cuts nor high-profile game releases were proving helpful to the Saturn's success.[96] Due to the Saturn's poor performance in North America, 60 of Sega of America's 200 employees were laid off in the fall of 1997.[92]

"I thought the Saturn was a mistake as far as hardware was concerned. The games were obviously terrific, but the hardware just wasn't there."

—Bernie Stolar, former president of Sega of America giving his assessment of the Saturn in 2009.[36]

As a result of the company's deteriorating financial situation, Nakayama resigned as president of Sega in January 1998 in favor of Irimajiri.[92] Stolar would subsequently accede to president of Sega of America.[96][117] Following five years of generally declining profits,[118] in the fiscal year ending March 31, 1998 Sega suffered its first parent and consolidated financial losses since its 1988 listing on the Tokyo Stock Exchange.[119] Due to a 54.8% decline in consumer product sales (including a 75.4% decline overseas), the company reported a net loss of ¥43.3 billion (US$327.8 million) and a consolidated net loss of ¥35.6 billion (US$269.8 million).[118] Shortly before announcing its financial losses, Sega revealed that it was discontinuing the Saturn in North America, with the goal of preparing for the launch of its successor.[92][96] Only 12 Saturn games were released in North America in 1998 (*Magic Knight Rayearth* being the final official release), compared to 119 in 1996.[120][121] The Saturn would last longer in Japan and Europe.[93] Rumors about the upcoming Dreamcast —spread mainly by Sega itself—were leaked to the public before the last Saturn games were released.[93] The Dreamcast was released on November 27, 1998 in Japan and on September 9, 1999 in North America.[122] The decision to abandon the Saturn effectively left the Western market without Sega games for over one year.[123] Sega suffered an additional ¥42.881 billion consolidated net loss in the fiscal year ending March 1999, and the company announced plans to eliminate 1,000 jobs, or nearly one-fourth of its workforce.[124][125]

By the end of 1998, Sega had sold an estimated 2.7 million Saturn units in North America, compared to 13.4 million PlayStation consoles sold by Sony.[120] 5 million Saturn units were sold in Japan by March 1998 (surpassing the Genesis' sales of 3.5 million in the country),[92][126] and 971,000 were sold in Europe by July 1998.[127] With lifetime sales estimated at 9.5 million units worldwide, the Saturn is considered a commercial failure.[128] Lack of distribution has been cited as a significant factor contributing to the Saturn's limited installation base, as the system's surprise launch damaged Sega's reputation with key retailers.[51] Conversely, Nintendo's long delay in releasing a 3D console and damage caused to Sega's reputation by poorly supported add-ons for the Genesis are considered major factors allowing Sony to gain a foothold in the market.[59][129]

### 1.1.2 Technical specifications

Featuring a total of eight processors[130] the Saturn's main central processing units are two Hitachi SH-2 microprocessors clocked at 28.6 MHz and capable of 56 MIPS.[17][51] The system contains a Motorola 68EC000 running at 11.3 MHz as a sound controller, a custom sound processor with an integrated Yamaha FH1[131] DSP running at 22.6 MHz[132] capable of up to 32 sound channels with both FM synthesis and 16-bit PCM sampling at a maximum rate of 44.1 kHz,[133] and two video display processors,[14] the VDP1 (which handles sprites, textures and polygons) and the VDP2 (which handles backgrounds).[132] Its double-speed CD-ROM drive is controlled by a dedicated Hitachi SH-1 processor to reduce load times.[34] The Saturn's System Control Unit (SCU), which controls all buses and functions as a co-processor of the main SH-2 CPU, has an internal DSP[17] running at 14.3 MHz.[132] The Saturn contains a cartridge slot for memory expansion,[130] 16 Mbit of work random-access memory (RAM), 12 Mbit of video RAM, 4 Mbit of RAM for sound functions, 4 Mbit of CD buffer RAM and 256 Kbit (32 KB) of battery backup RAM.[133] Its video output, provided by a stereo AV cable,[133] displays at resolutions from 320×224 to 704×224 pixels,[134] and is capable of displaying up to 16.77 million colors simultaneously.[133] Physically, the Saturn measures 260 mm × 230 mm × 83 mm (10.2 in × 9.1 in × 3.3 in). The Saturn was sold packaged with an instruction manual, one control pad, a stereo AV cable, and its 100V AC power supply, with a power consumption of approximately 15W.[133]

"One very fast central processor would be preferable. I don't think all programmers have the ability to program two CPUs—most can only get about one-and-a-half times the speed you can get from one SH-2. I think that only 1 in 100 programmers are good enough to get this kind of speed [nearly double] out of the Saturn."

—Yu Suzuki reflecting upon Saturn *Virtua Fighter* development.[17]

The Saturn had technically impressive hardware at the time of its release, but its complexity made harnessing this power difficult for developers accustomed to conventional programming.[135] The greatest disadvantage was that both CPUs shared the same bus and were unable to access system memory at the same time. Making full use of the 4 kB of cache memory in each CPU was critical to maintaining performance. For example, *Virtua Fighter* used one CPU for each character,[17] while *Nights* used one CPU for 3D environments and the other for 2D objects.[136] The Saturn's Visual Display Processor 2 (VDP2), which can generate and manipulate backgrounds,[137] has also been cited as one of the system's most important features.[19][78]

"The Sega Saturn couldn't do true 3D."

—Yukio Futatsugi, designer of *Panzer Dragoon II Zwei.*[138]

The Saturn's design elicited mixed commentary among game developers and journalists. Developers quoted by *Next Generation* in December 1995 described the Saturn as "a real coder's machine" for "those who love to get their teeth into assembly and really hack the hardware", with "more flexibility" and "more calculating power than the PlayStation". The Saturn's sound board was also widely praised.[19] By contrast, Lobotomy Software programmer Ezra Dreisbach described the Saturn as significantly slower than the PlayStation,[139] whereas Kenji Eno of WARP observed little difference between the two systems.[140] In particular, Dreisbach criticized the Saturn's use of quadrilaterals as its basic geometric primitive, in contrast to the triangles rendered by the PlayStation and the Nintendo 64.[139] Third-party development was initially hindered by the lack of useful software libraries and development tools, requiring developers to write in assembly language to achieve good performance. During early Saturn development, programming in assembly could offer a two-to-fivefold speed increase over C language.[17] The Saturn hardware is considered extremely difficult to emulate.[141] Sega responded to complaints about the difficulty of programming for the Saturn by writing new graphics libraries which were claimed to make development easier.[19] Sega of America also purchased a United Kingdom-based development firm, Cross Products, to produce the Saturn's official development system.[32][142] Despite these challenges, Treasure CEO Masato Maegawa stated that the Nintendo 64 was more difficult to develop for than the Saturn.[143] Traveller's Tales' Jon Burton opined that while the PlayStation was easier "to get started on ... you quickly reach [its] limits", whereas the Saturn's "complicated" hardware had the ability to "improve the speed and look of a game when all used together correctly."[144] A major point of criticism was the Saturn's use of 2D sprites to generate polygons and simulate 3D space. The PlayStation functioned in a similar manner, but also featured a dedicated "Geometry Transfer Engine" that rendered additional polygons. As a result, several analysts described the Saturn as an "essentially" 2D system.[6][17][145]

Several models of the Saturn were produced in Japan. An updated model in a recolored light gray was released in Japan at a price of ¥20,000 in order to reduce the system's cost.[146] Two models were released by third parties: Hitachi released a model known as the "Hi-Saturn" (a smaller Saturn model equipped with a car navigation function),[147] while JVC released the "V-Saturn".[133] Saturn controllers came in various color schemes to match different models of the console.[148] The system also supports several accessories. A wireless controller powered by AA batteries utilizes infrared signal to connect to the Saturn.[149] Designed to work with *Nights*, the Saturn 3D Pad is a fully functional controller that includes both a control pad and an analog stick for directional input.[150] Sega also released several versions of arcade sticks as peripherals, including the Virtua Stick,[151][152] the Virtua Stick Pro,[153] the Mission Analog Stick,[154] and the Twin Stick.[155] Sega also created a light gun peripheral known as the "Virtua Gun" for use with shooting games such as *Virtua Cop* and *The Guardian*,[156] as well as the Arcade Racer, a wheel for racing games.[157][158] The Play Cable allows for two Saturn consoles to be connected for multiplayer gaming across two screens,[159] while a multitap allows up to six players to play games on the same console.[160][161] The Saturn was designed to support up to 12 players on a single console, by using two multitaps.[162] RAM cartridges expand the amount of memory in the system.[163] Other accessories include a keyboard,[164] mouse,[165][166] floppy disk drive,[167] and movie card.[1][168]

Like the Genesis, the Saturn had an Internet-based gaming service. The Sega NetLink was a 28.8k modem that fit into the cartridge slot in the Saturn for direct dial multiplayer.[14] In Japan, a now defunct pay-to-play service was used.[169] It could also be used for web browsing and sending email. Because the NetLink was released before

the Saturn keyboard, Sega produced a series of CDs containing hundreds of website addresses so that Saturn owners could browse with the joypad.[170] The NetLink functioned with five games: *Daytona USA*, *Duke Nukem 3D*, *Saturn Bomberman*,[171] *Sega Rally*, and *Cyber Troopers Virtual-On: Operation Moongate*.[172] Sega allegedly developed a variant of the Saturn featuring a built-in NetLink modem under the code name "Sega Pluto", but it was never released.[173]

Sega developed an arcade board based on the Saturn's hardware, called the Sega ST-V (or Titan), which was intended as an affordable alternative to Sega's Model 2 arcade board as well as a testing ground for upcoming Saturn software.[17] The Titan was criticized for its comparatively weak performance by Sega AM2's Yu Suzuki[17] and was overproduced by Sega's arcade division.[106] Because Sega already possessed the *Die Hard* license, members of Sega AM1 working at the Sega Technical Institute developed *Die Hard Arcade* for the Titan, in order to clear out excess inventory.[106] This goal was achieved, as *Die Hard* became the most successful Sega arcade game produced in the United States at that point.[106] Other games released for the Titan include *Golden Axe: The Duel* and *Virtua Fighter Kids*.[17][69]

### 1.1.3 Game library

Main article: List of Sega Saturn games

Much of the Saturn's library comes from Sega's arcade ports,[36] including *Daytona USA*, *The House of the Dead*,[174] *Last Bronx*, *Sega Rally Championship*, the *Virtua Cop* series, the *Virtua Fighter* series, and *Virtual-On*.[175] The Saturn ports of 2D Capcom fighting games including *Darkstalkers 3*, *Marvel Super Heroes vs. Street Fighter*, and *Street Fighter Alpha 3* were noted for their faithfulness to their arcade counterparts.[175][176] *Fighters Megamix*, developed by Sega AM2 for the Saturn rather than arcades,[104] combined characters from *Fighting Vipers* and *Virtua Fighter* to positive reviews.[177] Highly rated Saturn exclusives include *Panzer Dragoon Saga*,[178] *Dragon Force*,[179] *Guardian Heroes*,[180][181] *Nights*,[182][183] *Panzer Dragoon II Zwei*,[184] and *Shining Force III*.[185][186][187] Although originally made for the PlayStation, games such as *Castlevania: Symphony of the Night*, *Resident Evil*, and *Wipeout 2097* received Saturn ports with mixed results.[175] *Tomb Raider* was created with the Saturn in mind, but the PlayStation version ultimately became better known to the public.[21][175][188] Lobotomy Software's *PowerSlave* featured some of the most impressive 3D graphics on the system, leading Sega to contract the developer to produce Saturn ports of *Duke Nukem 3D* and *Quake*.*[21]*[175] While Electronic Arts' limited support for the Saturn and Sega's failure to develop a football game for the 1996 fall season allowed Sony to take the lead in the sports genre,*[21]*[36]*[61] "Sega Sports" published Saturn sports games including the well-regarded *World Series Baseball* and *Sega Worldwide Soccer* series.*[21]*[189] With about 600 official releases, the Saturn's library is nearly twice as large as the Nintendo 64's.*[121]

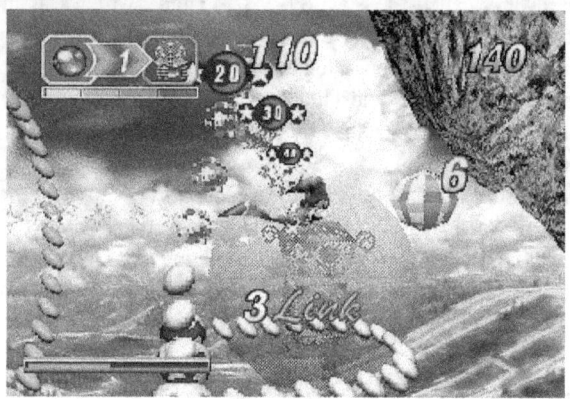

*A typical in-game screen shot of* NiGHTS into Dreams...*, taken from the "Splash Garden" level*

Due to the cancellation of *Sonic X-treme*, the Saturn lacks an exclusive *Sonic the Hedgehog* platformer, containing only the compilation *Sonic Jam*, a graphically enhanced port of the Genesis title *Sonic 3D Blast*, and a racing game called *Sonic R*.*[14]*[190] Notable Saturn platformers include *Bug!*, whose eponymous main character was considered to be a potential mascot.*[191] Despite receiving generally positive reviews at the time*[192] (and being successful enough to receive a sequel), *Bug!* failed to catch on with audiences in the way the *Sonic* series had, and the game garnered criticism due to its mostly 2D gameplay and perceived lack of originality.*[191]*[193] Considered one of the most important Saturn releases, Sonic Team developed *Nights into Dreams...*, a score attack game that attempted to simulate both the joy of flying and the fleeting sensation of dreams. The gameplay of *Nights* involves steering the imp-like androgynous protagonist, Nights, as it flies on a mostly 2D plane across surreal stages broken into four segments each. The levels repeat for as long as an-game time limit allows, while flying over or looping around various objects in rapid succession earns additional points. Although it lacked the fully 3D environments of Nintendo's *Super Mario 64*, *Nights'* emphasis on unfettered movement and graceful acrobatic techniques showcased the intuitive potential of analog control.*[136]*[194]*[195] Sonic Team's *Burning Rangers*, a fully 3D*[21] action-adventure game involving a team of outer-space firefighters, garnered praise

for its transparency effects and distinctive art direction, but was released in limited quantities late in the Saturn's lifespan and criticized for its short length.*[196]*[197]*[198]

Some of the games that made the Saturn popular in Japan, such as *Grandia*[21] and the *Sakura Wars* series, never saw a Western release due to Sega of America's policy of not localizing RPGs and other Japanese titles that might have damaged the system's reputation in North America.*[36]*[199] Despite appearing first on the Saturn, games such as *Dead or Alive*,*[175]*[200] *Grandia*,*[175] and *Lunar: Silver Star Story Complete* only saw a Western release on the PlayStation.*[21] Working Designs localized several Japanese Saturn games before a public feud between Sega of America's Bernie Stolar and Working Designs president Victor Ireland resulted in the company switching their support to the PlayStation.*[21] *Panzer Dragoon Saga* was praised as perhaps the finest RPG for the system due to its cinematic presentation, evocative plot, and unique battle system—with a tactical emphasis on circling around opponents to identify weak points and the ability to "morph" the physical attributes of the protagonist's dragon companion during combat—but Sega released fewer than 20,000 retail copies of the game in North America in what IGN's Levi Buchanan characterized as one example of the Saturn's "ignominious send-off" in the region.*[197]*[201]*[202] Similarly, only the first of three installments of *Shining Force III* was released outside Japan.*[187] The Saturn's library also garnered criticism for its lack of sequels to high-profile Genesis-era Sega franchises, with Sega of Japan's cancellation of a planned third installment in Sega of America's popular *Eternal Champions* series being cited as a significant source of controversy.*[21]*[121]*[203]

Later ports of Saturn games including *Guardian Heroes*,*[204] *Nights*,*[195] and *Shin Megami Tensei: Devil Summoner: Soul Hackers*[205] continued to garner positive reviews from critics. Partly due to rarity, Saturn titles such as *Panzer Dragoon Saga*[201]*[206]*[202] and *Radiant Silvergun*[207]*[208] have been noted for their cult following. Due to the system's commercial failure and hardware limitations, planned Saturn versions of games such as *Resident Evil 2*,*[209] *Shenmue*, *Sonic Adventure*, and *Virtua Fighter 3*[210]*[211] were cancelled and moved to the Dreamcast.

### 1.1.4   Reception and legacy

At the time of its release, *Famicom Tsūshin* awarded the Saturn console 24 out of 40 possible points, higher than the PlayStation's 19 out of 40.*[212] In June 1995, Dennis Lynch of the *Chicago Tribune* and Albert Kim of *Entertainment Weekly* both praised the Saturn as the most advanced gaming console available, with the former complimenting its double-speed CD-ROM drive and "intense surround-sound capabilities" and the latter citing *Panzer Dragoon* as a "lyrical and exhilarating epic" demonstrating the ability of new technology to "transform" the industry.*[213]*[214] In December 1995, *Next Generation* evaluated the system with three and a half stars out of a possible five, highlighting Sega's marketing and arcade background as strengths but the system's complexity as a weakness.*[19] *Electronic Gaming Monthly*'s December 1996 Buyer's Guide had four reviewers rate the Saturn 8, 6, 7, and 8 out of 10; these ratings were inferior to those of the PlayStation, which was scored 9, 10, 9, and 9 in the same review.*[215] By December 1998, *Electronic Gaming Monthly*'s reviews were more mixed, with reviewers citing the lack of titles for the system as a major issue. According to *EGM* reviewer Crispin Boyer, "the Saturn is the only system that can thrill me one month and totally disappoint me the next." *[216]

Retrospective feedback of the Saturn is mixed, but generally praises its game library.*[36]*[175] According to Greg Sewart of 1UP.com, "the Saturn will go down in history as one of the most troubled, and greatest, systems of all time." *[21] In 2009, IGN chose the Saturn to be their 18th best video game console of all time, praising its unique game library. According to the reviewers, "While the Saturn ended up losing the popularity contest to both Sony and Nintendo ... *NiGHTS into Dreams*, the *Virtua Fighter* and *Panzer Dragoon* series are all examples of exclusive titles that made the console a fan favorite." *[176] The staff of *Edge* noted "hardened loyalists continue to reminisce about the console that brought forth games like *Burning Rangers*, *Guardian Heroes*, *Dragon Force* and *Panzer Dragoon Saga*." *[217] In 2015, *The Guardian*'s Keith Stuart declared "the Saturn has perhaps the strongest line up of 2D shooters and fighting games in console history." *[218] *Retro Gamer*'s Damien McFerran stated "Even today, despite the widespread availability of sequels and re-releases on other formats, the Sega Saturn is still a worthwhile investment for those who appreciate the unique gameplay styles of the companies that supported it." *[14] IGN's Adam Redsell wrote "[Sega's] devil-may-care attitude towards game development in the Saturn and Dreamcast eras is something that we simply do not see outside of the indie scene today." *[172] Necrosoft Games director Brandon Sheffield expounded that "The Saturn was a landing point for games that were too 'adult' in content for other systems, as it was the only one that allowed an 18+ rating for content in Japan ... some games, like *Enemy Zero* used it to take body horror to new levels, an important step toward the expansion of games and who they served." *[121] Sewart praised the Saturn's first-party titles as "Sega's shining moment as a game developer", with Sonic Team demonstrating its creative range and AM2 producing numerous technically impressive arcade ports, but

also commented on the many Japan-exclusive Saturn releases, which he connected with a subsequent boom in the game import market.*[21] IGN's Travis Fahs was critical of the Saturn library's lack of "fresh ideas" and "precious few high-profile franchises" , in contrast to what he described as Sega's more creative Dreamcast output.*[123]

Criticism has befallen Sega's management regarding both the creation and handling of the Saturn. McFerran criticizes Sega's management at the time of the Saturn's development, claiming that they had "fallen out of touch with both the demands of the market and the industry" .*[14] Bernie Stolar has also been criticized for his decision to end support for the Saturn.*[21] According to Fahs, "Stolar's decision to abandon the Saturn made him a villain to many SEGA fans, but ... it was better to regroup than to enter the next fight battered and bruised. Dreamcast would be Stolar's redemption." *[36] Stolar has defended his decision, stating, "I felt Saturn was hurting the company more than helping it. That was a battle that we weren't going to win." *[96] Sheffield stated that the Saturn's use of quadrilaterals undermined third-party support for the system, but because "nVidia invested in quads" at the same time there is "a remote possibility" they could have "become the standard instead of triangles"—"if somehow, magically, the Saturn were the most popular console of that era." *[121] Speaking more positively of the system, former Working Designs president Victor Ireland described the Saturn as "the start of the future of console gaming" because it "got the better developers thinking and designing with parallel-processing architecture in mind for the first time" .*[21] Writing for GamesRadar, Justin Towell noted that the Saturn's 3D Pad "set the template for every successful controller that followed, with analog shoulder triggers and left thumbstick ... I don't see any three-pronged controllers around the office these days." *[219]

Douglass C. Perry of Gamasutra notes that, from its surprise launch to its ultimate failure, the Saturn "soured many gamers on Sega products." *[220] Sewart and IGN's Levi Buchanan cited the failure of the Saturn as the major reason for Sega's downfall as a hardware manufacturer, but USgamer's Jeremy Parish described the Saturn as "more a symptom ... than a cause" of the company's decline, which began with add-ons for the Genesis that fragmented the market and continued with Sega of America and Sega of Japan's competing designs for the Dreamcast.*[21]*[121]*[221] Sheffield portrayed Sega's mistakes with the Saturn as emblematic of the broader decline of the Japanese gaming industry: "They thought they were invincible, and that structure and hierarchy were necessary for their survival, but more flexibility, and a greater participation with the West could have saved them." *[121] According to Stuart, Sega "didn't see ... the roots of a prevailing trend, away from arcade conversions and traditional role-playing adventures and toward a much wider console development community with fresh ideas about gameplay and structure." *[222] Pulp365 reviews editor Matt Paprocki concluded "the Saturn is a relic, but an important one, which represents the harshness of progress and what it can leave in its wake". *[121]

## 1.1.5 References

[1] "Movie card" (in Japanese). Sega of Japan. Retrieved 2014-03-03.

[2] Sczepaniak, John (2006). "Retroinspection: Mega Drive". *Retro Gamer* (27): 42–47.

[3] Kent 2001, pp. 424, 427.

[4] Kent 2001, p. 428.

[5] Kent 2001, p. 431.

[6] McFerran, Damien (2010). "Retroinspection: Sega 32X". *Retro Gamer* (77): 44–49. **Scot Bayless:** The 32X call was made in early January [1994] ... There's a part of me that wishes the Saturn had adopted the 32X graphics strategy, but that ship had sailed long before the greenlight call from Nakayama.

[7] McFerran, Damien (2012-02-22). "The Rise and Fall of Sega Enterprises". *Eurogamer*. Retrieved 2014-05-01.

[8] "*Virtua Racing* – Arcade (1992)". *GameSpot*. 2001. Retrieved 2014-06-06. cf. Feit, Daniel (2012-09-05). "How *Virtua Fighter* Saved PlayStation's Bacon". *Wired*. Retrieved 2014-10-09. **Ryoji Akagawa:** If it wasn't for *Virtua Fighter*, the PlayStation probably would have had a completely different hardware concept. cf. Thomason, Steve (July 2006). "The Man Behind the Legend". *Nintendo Power* **19** (205): 72. **Toby Gard:** It became clear to me watching people play *Virtua Fighter*, which was kind of the first big 3D-character console game, that even though there were only two female characters in the lineup, in almost every game I saw being played, someone was picking one of the two females.

[9] Leone, Matt (2010). "The Essential 50 Part 35: *Virtua Fighter*". *1UP.com*. Archived from the original on 2013-10-19. Retrieved 2014-04-21.

[10] Donovan, Tristan (2010). *Replay: The History of Video Games*. Yellow Ant. p. 267. ISBN 978-0956507204. One of the key objections to 3D graphics that developers had been raising with Sony was that while polygons worked fine for inanimate objects such as racing cars, 2D images were superior when it came to animating people or other characters. *Virtua Fighter*, Suzuki's follow-up to *Virtua Racing*, was a direct riposte to such thinking ... The characters may have resembled artists' mannequins but their lifelike movement turned Suzuki's game into a huge success that exploded claims that game characters couldn't be done successfully in

3D ... Teruhisa Tokunaka, chief executive officer of Sony Computer Entertainment, even went so far as to thank Sega for creating *Virtua Fighter* and transforming developers' attitudes.

[11] Mott 2013, pp. 226, 250. "*Virtua Racing* ... was perhaps the first to treat polygons not as a graphical gimmick but as an opportunity to expand the boundaries of traditional driving games ... It's like witnessing the discovery of fire ... [*Virtua Fighter*] establish[ed] the template that future 3-D fighters would follow".

[12] Kent 2001, pp. 501–502.

[13] "*Virtua Fighter* Review". *Edge*. 1994-12-22. Retrieved 2015-03-05. *Virtua Fighter*'s 3D characters have a presence that 2D sprites just can't match. The characters really do seem 'alive', whether they're throwing a punch, unleashing a special move or reeling from a blow ... The Saturn version of *Virtua Fighter* is an exceptional game in many respects. It's arguably the first true 'next generation' console game, fusing the best aspects of combat gameplay with groundbreaking animation and gorgeous sound (CD music and clear samples). In the arcades, *Virtua Fighter* made people stop and look. On the Saturn, it will make many people stop, look at their bank balance and then fork out for Sega's new machine. Over to you, Sony.

[14] McFerran, Damien. "Retroinspection: Sega Saturn". *Retro Gamer* (34): 44–49.

[15] Harris 2014, p. 386.

[16] "EGM Interviews SEGA SATURN Product Manager HIDEKI OKAMURA". *EGM²* **1** (1). July 1994. **Hideki Okamura:** [Saturn] was just a development code name for hardware that was adopted by the Japanese development staff. The name has become common knowledge and it has a nice ring to it.

[17] "Sega Saturn". *Next Generation* **1** (2): 36–43. February 1995. Sega's knee-jerk reaction was to delay it's Saturn development program for a few months to incorporate a new video processor into the system. Not only would this boost its 2D abilities considerably (something that Sony's machine was less proficient at), but it would also provide better texture mapping for 3D graphics ... Of course, Hitachi's link with the Saturn project goes much deeper. In 1993, the Japanese electronics company set up a joint venture with Sega to develop a CPU for the Saturn based on proprietary Hitachi technology. Several Hitachi staff were seconded to Sega's Saturn division (it's now believed that the same team is now working on preliminary 64-bit technology for Sega), and the result was the SH-2 ... As with most Sega hardware, Model 1 was basically an expensive assortment of bought-in chips. Its main CPU, an NEC V60 running at just 16 MHz, was simply too slow for the Saturn. And the bulk of *Virtua Racing*'s number crunching was handled by four serial DSPs that were way too costly to be included in any home system. Sega's consequent development of the SH-2 meant that it could also produce a Saturn-compatible arcade system.

[18] Pollack, Andrew (1993-09-22). "Sega to Use Hitachi Chip In Video Game Machine". *The New York Times*. Retrieved 2014-04-15. Sega Enterprises said today that it would base its next-generation home video game machine, due in the fall of 1994, on a new chip being developed by Hitachi Ltd ... One Sega official said Hitachi's chip was attractively priced and would be designed with Sega's needs in mind ... Yamaha is expected to provide sound chips and JVC the circuitry for compressing video images. cf. "Sega to add 64-Bit Processor to New Saturn System!". *Electronic Gaming Monthly* **5** (53): 68. December 1993. There are reportedly seven different processors in the Saturn. The main processor will be a custom 32-Bit RISC chip under joint development by Sega and Hitachi.

[19] "NG Hardware: Saturn". *Next Generation* **1** (12): 45–48. December 1995. The early pictures and technical breakdowns have remained relatively close to the final system, perhaps because the system was completed far earlier than many people realize ... It was too late to make major alterations to the system, so, at the cost of pushing the launch schedule slightly, a video processor was added to the board to boost its 2D and 3D texture-mapping abilities. The real processing power of the Saturn comes from two Hitachi SH2 32-bit RISC processors running at 28 MHz. These processors were specially commissioned by Sega and are optimized for fast 3D graphics work.

[20] "NG Hardware: Saturn". *Next Generation* **1** (1): 44–45. January 1995. Sega has spent the last nine months or so playing catch-up with Sony after a publisher-friend tipped Sega off about the power of PlayStation.

[21] Sewart, Greg (2005-08-05). "Sega Saturn: The Pleasure And The Pain". *1UP.com*. Retrieved 2014-11-27.

[22] Fahs, Travis (2009-04-21). "IGN Presents the History of Sega". *IGN*. p. 6. Retrieved 2014-05-01.

[23] Dring, Christopher (2013-07-07). "A Tale of Two E3s – Xbox vs Sony vs Sega". *MCVUK.com*. Retrieved 2014-03-19.

[24] Harris 2014, p. 465.

[25] Harris 2014, p. 464.

[26] Horowitz, Ken (2006-07-11). "Interview: Tom Kalinske". *Sega-16*. Retrieved 2014-12-24. **Tom Kalinske:** I remember we had a document that Olaf and Mickey took to Sony that said they'd like to develop jointly the next hardware, the next game platform, with Sega, and here's what we think it ought to do. Sony apparently gave the green light to that ... Our proposal was that each of us would sell this joint Sega/Sony hardware platform; we'll share the loss on the hardware (whatever that is, we'll split it), combine our advertising and marketing, but we'll each be responsible for the software sales we'll generate. Now, at that particular point in time, Sega knew how to develop software a hell of a lot better than Sony did. They were just coming up the learning curve, so we would have benefited much more greatly ... I felt that we were rushing Saturn. We didn't have the software right, and we didn't have the pricing right, so I felt we should have stayed with Genesis for another year.

[27] Harris 2014, p. 452.

[28] Kent 2001, p. 509.

[29] "The Making Of ... *Panzer Dragoon Saga* Part 1". *Now Gamer*. 2008-12-17. Retrieved 2014-03-20. **Kentaro Yoshida:** We thought we'd have no problem making games that were superior to PlayStation games.

[30] Kent 2001, p. 494.

[31] Beuscher, David. "Sega Genesis 32X – Overview". *Allgame*. Retrieved 2014-12-13.

[32] Horowitz, Ken (2013-02-07). "Interview: Joe Miller". *Sega-16*. Retrieved 2014-05-25. **Joe Miller:** I'd say that the rhetoric around the deteriorating relationship is probably overblown a little bit, based on what I've read. Nakayama-san and SOJ knew they had a strong, proven management team in place at SOA, and while everyone was concerned about growing the business, neither side lost confidence in the other.

[33] "Sega Saturn" (in Japanese). Sega of Japan. Retrieved 2014-03-03.

[34] "Sega and Sony Sell the Dream". *Edge* **3** (17): 6–9. February 1995. The December 3 ship-out of 100,000 PlayStations to stores across Japan ... was not met with the same euphoria-charged reception that the Saturn received ... Saturn arrived to a rapturous reception in Japan on November 22. 200,000 units sold out instantly on day one ... Japanese gamers were beside themselves as they walked away with their prized possession and a near-perfect conversion of the *Virtua Fighter* coin-op ... Sega (and Sony) have proved that with dedicated processors handling the drive (the SH-1 in the Saturn's case), negligible access times are possible.

[35] Semrad, Ed (December 1994). "Saturn... Ahead of its Time?". *Electronic Gaming Monthly* (65). p. 6.

[36] Fahs, Travis (2009-04-21). "IGN Presents the History of Sega". *IGN*. p. 8. Retrieved 2014-05-01.

[37] Harris 2014, p. 536, gives a lower figure of 170,000.

[38] Kent 2001, p. 502.

[39] Buchanan, Levi (2008-10-24). "32X Follies". *IGN*. Retrieved 2013-05-25.

[40] "Super 32X" (in Japanese). Sega of Japan. Retrieved 2014-02-23.

[41] "The Making Of: PlayStation". *Edge*. 2009-04-24. p. 4. Retrieved 2015-03-05.

[42] "Sega Saturn: You've Watched the TV Commercials...Now Read the Facts". *Next Generation* **1** (8): 26–32. August 1995.

[43] "History of the PlayStation". *IGN*. Retrieved 2014-11-16.

[44] Kent 2001, p. 504.

[45] "The Making Of: PlayStation". *Edge*. 2009-04-24. p. 3. Retrieved 2015-03-05.

[46] Kent 2001, p. 516.

[47] "Let the games begin: Sega Saturn hits retail shelves across the nation Sept. 2; Japanese sales already put Sega on top of the charts." . *Business Wire*. 1995-03-09. Retrieved 2014-12-24.

[48] Harris 2014, p. 536.

[49] "A Saturn Surprise!". *GamePro* **7** (73): 30. August 1995.

[50] Cifaldi, Frank (2010-05-11). "This Day in History: Sega Announces Surprise Saturn Launch". *1UP.com*. Archived from the original on 2014-05-28. Retrieved 2014-05-04.

[51] Schilling, Mellissa A. (Spring 2003). "Technological Leapfrogging: Lessons From the U.S. Video Game Console Industry". *California Management Review* **45** (3): 12, 23. Lack of distribution may have contributed significantly to the failure of the Sega Saturn to gain an installed base. Sega had limited distribution for its Saturn launch, which may have slowed the building of its installed base both directly (because consumers had limited access to the product) and indirectly (because distributors that were initially denied product may have been reluctant to promote the product after the limitations were lifted). Nintendo, by contrast, had unlimited distribution for its Nintendo 64 launch, and Sony not only had unlimited distribution, but had extensive experience with negotiating with retailing giants such as Wal-Mart for its consumer electronics products.

[52] cf. "Is War hell for Sega?". *Next Generation* **2** (13): 7. January 1996. **Tom Kalinske:** We needed to do something shocking because we were $100 more than the other guy ... I still think [the surprise launch] was a good idea. If I had it to do over again would I do it a little differently? Yeah, definitely. I wouldn't take the risk of annoying retailers the way we did. I would clue them in and do an early launch in a region or three regions or something so we could include everybody.

[53] Harris 2014, p. 545.

[54] Kent 2001, pp. 505, 516.

[55] "Dear Saturn Mag, I've Heard the Saturn Couldn't Handle *Alex Kidd...* Is This True?". *Sega Saturn Magazine* **1** (2). December 1995. p. 51.

[56] Horsman, Mathew (1995-11-11). "Sega profits plunge as rivals turn up the heat". *The Independent*. Retrieved 2015-01-20.

[57] "Sony Computer Entertainment Inc. Business Development/Europe". SCE. Retrieved 2015-01-20.

[58] "Sega Saturn gets astronomical send off with landmark marketing campaign; Sega breaks $50-million marketing campaign to support surprise launch at E3." . *Business Wire*. 1995-05-11. Retrieved 2015-02-18.

[59] Mäyrä, Frans (editor); Finn, Mark (2002). "Console Games in the Age of Convergence". *Computer Games and Digital Cultures: Conference Proceedings: Proceedings of the Computer Games and Digital Cultures Conference, June 6–8, 2002, Tampere, Finland*. Tampere University Press. pp. 45–58. ISBN 9789514453717.

[60] Kato, Matthew (2013-10-30). "Which Game Console Had The Best Launch Lineup?". *Game Informer*. p. 3. Retrieved 2014-03-19.

[61] Kent 2001, p. 533.

[62] DeMaria & Wilson 2004, p. 282.

[63] Kent 2001, pp. 519–520.

[64] Parkin, Simon (2014-06-19). "A History of Videogame Hardware: Sony PlayStation". *Edge*. Retrieved 2015-03-05.

[65] "*Daytona USA*". *Edge* **3** (21): 72–5. June 1995. Although AM2 has managed to replicate the coin-op tolerably well, Saturn *Daytona* fails to capture the arcade experience that PlayStation *Ridge Racer* so convincingly delivers. cf. McNamara, Andy et al. (September 1995). "Prepare Yourself for the Ultimate Racing Experience". *Game Informer*. Archived from the original on 1997-11-20. Retrieved 2014-04-15. *Daytona* rules the arcade, but I think *Ridge Racer* dominates the home systems. cf. Air Hendrix (August 1995). "Pro Review: *Daytona USA*". *GamePro* **7** (73): 50. *Daytona* pales in comparison to *Ridge Racer* for the Japanese PlayStation, which takes an early lead with better features, gameplay, and graphics.

[66] Mott 2013, p. 239. "A disastrous home version [of *Daytona USA*] for the Sega Saturn in 1995 is reviled for its choppy frame rate and flickering polygons" .

[67] Kent 2001, p. 582.

[68] "*Tekken*". *Edge* **3** (21): 66–70. June 1995. Namco took a significant risk in basing its *Tekken* coin-op on raw PlayStation hardware, considering that it would be competing directly with Sega's Model 2-powered *Virtua Fighter 2* ... For once, a home system can boast an identical conversion of a cutting-edge coin-op ... Namco's research section managing director, Shegeichi Nakamura ... explains: "When Sony came along we decided to go for a low-cost system—in short, we've left the big arcade stores to Sega and *VF2* and *Tekken* has been sold to smaller arcade centres" ... Namco has a further four titles planned for System 11, all of which are likely to make the jump to the PlayStation.

[69] Tokyo Drifter (April 2002). "Virtua Fight Club". *GamePro* **14** (163): 48–50.

[70] "Namco". *Next Generation* **1** (1): 70–73. January 1995.

[71] *"Tekken"*. *Next Generation* **1** (2): 82. February 1995.

[72] "An Audience With: Katsuhiro Harada – on 20 years of *Tekken* and the future of fighting games". *Edge*. 2013-09-23. Retrieved 2015-03-05.

[73] Mott 2013, p. 254.

[74] cf. Scary Larry (August 1995). "Pro Review: *Virtua Fighter*". *GamePro* **7** (73): 48. The graphics were state-of-the-art when this game was released in the arcades a year ago. Other fighters—notably *Tekken* and *Toh Shin Den*—now make better use of the polygon engine.

[75] "Sega announces $299 Sega Saturn core pack; *"Virtua Fighter Remix"* pack-in available for $349.". *Business Wire*. 1995-10-02. Retrieved 2014-12-24. Sega of America Monday announced that, effective immediately, it will dramatically drop the price of its high-end Sega Saturn system to $299.

[76] cf. Reiner, Andrew et al. (January 1996). "Easy Left, Baby". *Game Informer*. Retrieved 2014-09-16. I'm far more impressed with this title than I was with Daytona. cf. "Top Gear". *Next Generation* **2** (14): 160. February 1996.

[77] cf. Reiner, Andrew et al. (January 1996). "Rendered and Ready to Wear". *Game Informer*. Retrieved 2014-09-16. cf. "Stunning". *Next Generation* **2** (14): 162. February 1996. Totally eliminates the hit or miss polarity of other light-gun games and adds a whole new level of detail to the genre.

[78] *"Virtua Fighter 2* is Here at Last!". *Next Generation*. Archived from the original on 1997-04-19. Retrieved 2014-04-12. [The VDP2] can generate and manipulate 3D backgrounds. This leaves the twin processors free to deal with manipulating the fighters themselves. The result is swift, elegant animation at 60 frames a second—the same speed as the *VF2* coin-op ... Sony's machine does not have an equivalent of the VDP2, so the demands for better animation and more realistic movement are placing greater and greater pressure on its central processor.

[79] Marriott, Scott Alan. *"Virtua Fighter 2"*. *Allgame*. Retrieved 2014-12-14.

[80] cf. "Platinum Pick: Virtua Fighter 2". *Next Generation* **2** (13): 179. January 1996. The ultimate arcade translation ... the best fighting game ever. cf. "Excellent!". *Next Generation* **2** (14): 160. February 1996. A general attention to detail that sets a new mark for quality game design.

[81] "Sony fights Sega on US streets". *Next Generation* **2** (13): 14–16. January 1996.

[82] "Sega captures dollar share of videogame market again; diverse product strategy yields market growth; Sega charts path for 1996.". *Business Wire*. 1996-01-10. Retrieved 2014-12-24. Estimated dollar share for Sega-branded interactive entertainment hardware and software in 1995 was 43 percent, compared with Nintendo at 42 percent, Sony at 13 percent and The 3DO Co. at 2 percent. Sega estimates the North American videogame market will total more than $3.9 billion for 1995.

[83] Kent 2001, p. 532.

[84] Kent 2001, p. 531.

[85] Gallagher, Scott; Park, Seung Ho (February 2002). "Innovation and Competition in Standard-Based Industries: A Historical Analysis of the U.S. Home Video Game Market". *IEEE Transactions on Engineering Management* **49** (1): 67–82.

[86] Kent 2001, p. 508.

[87] Kent 2001, p. 535. **Michael Latham:** "[Tom] would fall asleep on occasion in meetings. That is true. These were nine-hour meetings. Sega had a thing for meetings. You'd get there at 8:00 A.M. and then you'd get out of the meeting at, like, 4:00 P.M., so he wasn't the only person ... It wasn't the failure of the Saturn that made him lose interest; it was the inability to do something about it. He was not allowed to do anything. The U.S. side was basically no longer in control".

[88] Kent 2001, p. 534.

[89] "NEWSFLASH: Sega Planning Drastic Management Reshuffle – World Exclusive". *Next Generation*. 1996-07-13. Retrieved 2014-05-06.

[90] "Sega of America appoints Shoichiro Irimajiri chairman/chief executive officer". *M2PressWIRE*. 1996-07-16. Retrieved 2014-12-24. Sega of America Inc. (SOA) Monday announced that Shoichiro Irimajiri has been appointed chairman and chief executive officer. Sega also announced that Bernard Stolar, previously of Sony Computer Entertainment America, has joined the company as executive vice president, responsible for product development and third-party business ... Sega also announced that Hayao Nakayama and David Rosen have resigned as chairman and co-chairman of Sega of America, respectively. (Subscription required.)

[91] "Kalinske Out – WORLD EXCLUSIVE". *Next Generation*. 1996-07-16. Retrieved 2014-05-06.

[92] Stephanie Strom (1998-03-14). "Sega Enterprises Pulls Its Saturn Video Console From the U.S. Market". *The New York Times*. Retrieved 2014-12-07.

[93] Kent 2001, p. 559.

[94] "Irimajiri Settles In At Sega". *Next Generation*. 1996-07-25. Retrieved 2014-05-06. Although a familiar face at Sega of America, Shoichiro Irimajiri has spent his first week in charge re-meeting all the staff.

[95] Kent 2001, p. 535.

[96] Kent 2001, p. 558.

[97] Kent 2001, p. 506.

[98] Johnston, Chris (1998-07-15). "Stolar Talks Dreamcast". *GameSpot*. Retrieved 2014-12-17. **Bernie Stolar:** I'm also a big believer in RPGs as well. No one ever believes that because I came out of the coin-op side of the business. But I'm an older, wiser person these days.

[99] Towell, Justin (2012-06-23). "'Mr. Sega Saturn' lives on via amazing T-shirt". *GamesRadar*. Retrieved 2014-03-03.

[100] "This Week in Japan". *Edge*. 2008-06-06. Retrieved 2015-03-05.

[101] Kolan, Patrick (2007-08-07). "*Shenmue*: Through the Ages". *IGN*. Retrieved 2014-04-30.

[102] "The Making of *Shenmue*: Yu Suzuki on the Cult Classic's Genesis, Development—And Its Future". *Edge*. 2014-03-20. Retrieved 2015-03-05.

[103] Corriea, Alexa Ray (2014-03-19). "Creator Yu Suzuki shares the story of *Shenmue*'s development". *Polygon*. Retrieved 2014-12-15.

[104] "Shenmue, the History: Our look at Shenmue's history begins back in 1996". *IGN*. 1999-07-13. Retrieved 2014-04-30.

[105] Mott 2013, p. 406.

[106] Horowitz, Ken (2007-06-11). "Developer's Den: Sega Technical Institute". *Sega-16*. Retrieved 2014-04-16. **Roger Hector:** When it became obvious that Sony was taking the lead, Sega's corporate personality changed. It became very political, with lots of finger-pointing around the company. Sega tried to get a handle on the situation, but they made a lot of mistakes, and ultimately STI was swallowed up in the corporate turmoil.

[107] Fahs, Travis (2008-05-29). "*Sonic X-Treme* Revisited – Saturn Feature at IGN". *IGN*. Retrieved 2014-04-30.

[108] Houghton, David (2008-04-24). "The greatest *Sonic* game we never got ...". *GamesRadar*. Retrieved 2012-07-23.

[109] "The Making Of... *Sonic X-treme*". *Edge* **15** (177): 100–103. July 2007.

[110] Barnholt, Ray. "Yuji Naka Interview: *Ivy the Kiwi* and a Little Sega Time Traveling". *1UP.com*. Archived from the original on 2014-09-07. Retrieved 2014-03-04.

[111] Towell, Justin (2012-06-23). "Super-rare 1990 *Sonic The Hedgehog* prototype is missing". *GamesRadar*. Retrieved 2014-03-04. **Yuji Naka:** The reason why there wasn't a Sonic game on Saturn was really because we were concentrating on NiGHTS. We were also working on *Sonic Adventure*—that was originally intended to be out on Saturn, but because Sega as a company was bringing out a new piece of hardware—the Dreamcast—we resorted to switching it over to the Dreamcast, which was the newest hardware at the time. So that's why there wasn't a Sonic game on Saturn.

With regards to *X-treme*, I'm not really sure on the exact details of why it was cut short, but from looking at how it was going, it wasn't looking very good from my perspective. So I felt relief when I heard it was cancelled.

[112] Kent 2001, p. 500.

[113] Kent 2001, p. 520.

[114] McCarthy, Dave (2006-11-20). "PlayStation-The total history". *Eurogamer*. Retrieved 2014-11-16.

[115] "The Making Of: *Final Fantasy VII*". *Edge*. 2012-08-26. p. 3. Retrieved 2015-03-05.

[116] Mott 2013, p. 332.

[117] Feldman, Curt (1998-04-22). "Katana Strategy Still on Back Burner". *GameSpot*. Retrieved 2014-12-09.

[118] "Sega Enterprises Annual Report 1998" (PDF). Sega Enterprises, Ltd. pp. 1, 7–8. Retrieved 2014-12-07.

[119] "Sega News From Japan". *GameSpot*. 1998-03-18. Retrieved 2014-12-07.

[120] Lemos, Robert (1999-03-17). "Sega makes play for Dreamcast support". *ZDNet*. Retrieved 2014-12-17.

[121] Parish, Jeremy (2014-11-18). "The Lost Child of a House Divided: A Sega Saturn Retrospective". *USgamer*. Retrieved 2014-12-17.

[122] Kent 2001, pp. 563–564.

[123] Fahs, Travis (2010-09-09). "IGN Presents the History of Dreamcast". *IGN*. Retrieved 2014-12-24.

[124] "Sega Corporation Annual Report 2000" (PDF). Sega Corporation. p. 18. Retrieved 2014-12-24.

[125] King, Sharon R. (1999-07-12). "TECHNOLOGY; Sega Is Giving New Product Special Push". *The New York Times*. Retrieved 2014-12-24.

[126] "Videospiel-Algebra". *Man!ac Magazine* (in German). May 1995.

[127] "Advanced Consoles". *Screen Digest 1998*. Screen Digest Ltd. July 1998.

[128] Lefton, Terry (1998). "Looking for a Sonic Boom". *Brandweek* **9** (39): 26–29.

[129] DeMaria & Wilson 2004, pp. 282–283.

[130] Beuscher, Dave. "Sega Saturn – Overview". *Allgame*. Retrieved 2014-12-13.

[131] Day, Rebecca (December 1996). "Battle of the Games". *Popular Mechanics* **173** (12): 52.

[132] "Saturn Overview Manual". Sega of America. 1994-06-06.

[133] "Sega Saturn various data" (in Japanese). Sega of Japan. Retrieved 2014-02-27.

[134] "VDP1 User Manual". Sega of America. 1995-06-27. p. 134.

[135] Kent 2001, p. 509. "In theory, Saturn, which featured two Hitachi SH2 32-bit central processing chips, was more powerful than PlayStation. The truth was that the SH2 chips were somewhat inferior to the chip Sony had selected ... and allotting different operations to both of the processing chips proved nearly impossible".

[136] "*Nights into Dreams* (review)". *Edge*. 1996-08-02. Retrieved 2015-03-05. cf. "*Nights Into Dreams* Retrospective". *Edge*. 2007-06-08. Retrieved 2015-03-05. cf. "Retrospective: *Nights Into Dreams*". *Edge*. 2014-03-15. Retrieved 2015-03-05. The 3D environments were drawn by one processor, while another handled the 2D enemies, hoops and trees, melding them seamlessly to create a smooth, surprisingly fast-moving game that still looks striking today.

[137] "Saturn Technical Specs". *Next Generation*. Archived from the original on 1996-12-20. Retrieved 2014-04-22.

[138] Mielke, James (2007-09-11). "*Panzer Dragoon Zwei* Retrospective". *1UP.com*. p. 2. Retrieved 2014-11-17.

[139] "Interview: Ezra Dreisbach". *Curmudgeon Gamer*. 2002-07-09. Archived from the original on 2007-09-27. Retrieved 2014-12-24. **Ezra Dreisbach:** And really, if you couldn't tell from the games, the PSX is way better than the Saturn. It's way simpler and way faster. There are a lot of things about the Saturn that are totally dumb. Chief among these is that you can't draw triangles, only quadrilaterals.

[140] Bettenhausen, Shane; Mielke, James. "Kenji Eno: Reclusive Japanese Game Creator Breaks His Silence". *1UP.com*. Archived from the original on 2013-11-03. Retrieved 2014-03-22. **Kenji Eno:** But, the PlayStation and the Saturn aren't that different, so moving it [*Enemy Zero*] to Saturn wasn't too difficult.

[141] Moss, Richard (2014-06-02). "Life after Death: Meet the People Ensuring that Yesterday's Systems Will Never be Forgotten". *Edge*. Retrieved 2015-03-05. Hackers are still unsure how some components work.

[142] "The official development system". *Edge* **3** (23): 55. August 1995.

[143] "Treasure Talks Yuke Yuke". *IGN*. 1997-04-14. Retrieved 2014-05-26.

[144] "Traveller's Tales: *Sonic R* Programmer Speaks!". *Sega Saturn Magazine* **3** (24): 25. October 1997.

[145] "Inside the PlayStation". *Next Generation* **1** (6): 51. June 1995.

[146] "Sega Saturn HST-0014" (in Japanese). Sega of Japan. Retrieved 2014-03-03.

[147] "On the Move!". *Sega Saturn Magazine* **2** (4). February 1996. p. 9.

[148] "Sega Saturn controller" (in Japanese). Sega of Japan. Retrieved 2014-03-03.

[149] "Sega Saturn wireless controller" (in Japanese). Sega of Japan. Retrieved 2014-03-03.

[150] "Sega Saturn Multi-controller" (in Japanese). Sega of Japan. Retrieved 2014-03-03.

[151] "Virtua Stick" (in Japanese). Sega of Japan. Retrieved 2014-03-03.

[152] "Stick Fix". *GamePro* (IDG) (84): 118. September 1995.

[153] "Virtua Stick Pro" (in Japanese). Sega of Japan. Retrieved 2014-03-03.

[154] "Mission analog stick" (in Japanese). Sega of Japan. Retrieved 2014-03-03.

[155] "Twin stick" (in Japanese). Sega of Japan. Retrieved 2014-03-03.

[156] "Virtua Gun" (in Japanese). Sega of Japan. Retrieved 2014-03-03.

[157] "Racing controller" (in Japanese). Sega of Japan. Retrieved 2014-03-03.

[158] "Sega's Saturn is off to the Races!". *Electronic Gaming Monthly* **7** (73): 30. August 1995.

[159] "Play cable" (in Japanese). Sega of Japan. Retrieved 2014-03-03.

[160] "Six Packed". *GamePro* (IDG) (85): 154. October 1995.

[161] "Multi-Terminal 6" (in Japanese). Sega of Japan. Retrieved 2014-03-03.

[162] "Sega's Saturn Launched in Japan". *Electronic Gaming Monthly* (65). December 1994. p. 60.

[163] "RAM cartridge" (in Japanese). Sega of Japan. Retrieved 2014-03-03.

[164] "Sega Saturn keyboard" (in Japanese). Sega of Japan. Retrieved 2014-03-03.

[165] "Shuttle mouse" (in Japanese). Sega of Japan. Retrieved 2014-03-03.

[166] "The Sega Saturn Enters Orbit". *GamePro* (68). March 1995. p. 30.

[167] "Sega Saturn floppy disk drive" (in Japanese). Sega of Japan. Retrieved 2014-03-03.

[168] "Bring the Noise". *Sega Saturn Magazine* **1** (1). November 1995. pp. 56–57.

[169] "Sega Saturn modem" (in Japanese). Sega of Japan. Retrieved 2014-03-03.

[170] "Saturn to Get Internet Connection Facilities in '96!". *Sega Saturn Magazine* **2** (5). March 1996. p. 8.

[171] Mott 2013, p. 309.

[172] Redsell, Adam (May 20, 2012). "SEGA: A Soothsayer of the Games Industry". *IGN*. Retrieved 2014-03-03.

[173] Blagdon, Jeff (2013-04-17). "Forgotten Sega Pluto console prototype surfaces online (update)". *The Verge*. Retrieved 2014-03-22.

[174] House, Michael L. "*The House of the Dead* Review". *All-game*. Retrieved 2014-12-13.

[175] "Best Saturn games of all time". *GamesRadar*. 2014-03-06. Retrieved 2014-04-06. But that doesn't mean it's a total bust. Numerous excellent games were released for the console, which was supported primarily in the mid-to-late 1990s, including a variety of original Sega classics and several stellar third-party releases. RPG and fighting game fans, in particular, enjoyed a healthy array of options on the platform.

[176] "SEGA Saturn is number 18". *IGN*. Retrieved 2012-01-23.

[177] "*Fighters Megamix* for Saturn". *GameRankings*. Retrieved 2014-03-26. cf. McNamara, Andy et al. (May 1997). "*Fighters Megamix – Saturn*". *Game Informer*. Archived from the original on 1999-08-24. Retrieved 2014-03-19. This has to be one of the finest fighters to ever grace consoles. cf. Williamson, Colin. "*Fighters Megamix* Review". *Allgame*. Retrieved 2014-12-13.

[178] "*Panzer Dragoon Saga* for Saturn". *GameRankings*. Retrieved 2014-03-26. cf."*Panzer Dragoon Saga*". *Game Informer*. May 1998. Archived from the original on 1999-08-24. Retrieved 2014-03-26. Only *Final Fantasy VII* tops it. cf. "*Panzer Dragoon Saga* Review". *Edge*. 1998-03-25. Retrieved 2015-03-05. Adds a creative depth only Square-Soft games can currently rival.

[179] "*Dragon Force* for Saturn". *GameRankings*. Retrieved 2014-03-26.

[180] "*Guardian Heroes* for Saturn". *GameRankings*. Retrieved 2014-03-26.

[181] Mott 2013, p. 300.

[182] "*NiGHTS Into Dreams...* for Saturn". *GameRankings*. Retrieved 2014-03-26.

[183] Mott 2013, p. 302.

[184] "*Panzer Dragoon II Zwei* for Saturn". *GameRankings*. Retrieved 2014-03-26.

[185] "*Shining Force III* for Saturn". *GameRankings*. Retrieved 2014-03-26.

[186] *Shining* creators Hiroyuki Takahashi and Shugo Takahashi have named *Shining the Holy Ark* and *Shining Force 3* their favorite games in the series. See "Power Profiles: Takahashi Brothers". *Nintendo Power* 21 (229): 80–83. August 2008.

[187] Mott 2013, p. 350.

[188] Blache III, Fabian; Fielder, Lauren. "The History of *Tomb Raider*: Series History". *GameSpot*. Retrieved 2014-06-06.

[189] cf. "Sega Sports Does It One More Time". *Game Informer*. November 1995. Archived from the original on 1999-08-24. Retrieved 2014-03-19. *World Series Baseball* is by far the smoothest baseball game ever made. cf. "*Worldwide Soccer '98*". *Game Informer*. January 1998. Archived from the original on 1999-09-30. Retrieved 2014-03-19. The graphics are smooth, and the physics are perfect.

[190] "*Sonic 3D Blast*". *Game Informer*. January 1997. Retrieved 2014-11-27.

[191] Buchanan, Levi (2009-02-02). "What Hath *Sonic* Wrought? Vol. 10". *IGN*. Retrieved 2014-03-15. **Steven Spielberg, CES 1995:** This is the character! This is the character that is going to do it for Saturn!

[192] "*Bug!* for Saturn". *GameRankings*. Retrieved 2014-03-15. cf. McNamara, Andy et al. (September 1995). "Not To Be Denied!". *Game Informer*. Retrieved 2014-03-15. cf. "*Bug!* (review)". *Electronic Gaming Monthly* 7 (73): 38. August 1995.

[193] "*Bug!*". *Next Generation* 1 (9): 88–89. September 1995.

[194] Buchanan, Levi (2008-08-18). "*NiGHTS into Dreams...* Retro Review". *IGN*. Retrieved 2014-12-17.

[195] Robinson, Martin (2012-05-10). "*NiGHTS Into Dreams* HD review". *Eurogamer*. Retrieved 2014-12-17.

[196] "Classic Reviews: *Burning Rangers*". *Game Informer* 12 (110): 104. June 2002. This futuristic fire-fighting game was an instant cult classic ... The game has all the makings of a masterpiece, but is held back by both the Saturn's limited power and the fireman motif.

[197] Buchanan, Levi (2008-09-03). "*Burning Rangers* Retro Review". *IGN*. Retrieved 2014-11-18. A wholly competent (but somewhat workmanlike, as the Saturn was not as good at 3D as the PSone or N64) action-adventure game.

[198] Mott 2013, p. 353.

[199] cf. "Japan Votes on All Time Top 100". *Edge*. 2006-03-03. Retrieved 2015-03-05.

[200] cf. "Catching up with Tecmo's Prince of Darkness". *Game Informer* 15 (140): 203. December 2004. **Tomonobu Itagaki:** Saturn was a great machine, great system. If someone were to ask me if I want to make a game for PSone or Saturn I would—100 percent of the time—make games for the Saturn.

[201] "*Panzer Dragoon Saga*: The Sad Tale of the Saturn's Last Great Game". *Game Informer* 17 (176): 164–165. December 2007. One of the greatest games ever crafted by human hands ... Critically, the game was a smash hit, lauded as one of the year's best, and generally considered the Saturn's finest title. But despite glowing reviews across the board, *Saga* was

destined to fail. Sega had moved on—shifted its focus to developing its next console, the Dreamcast, and wasn't willing to risk any more money on a system that had already lost so much. Therefore, less than 20,000 retail copies of *Panzer Dragoon Saga* were even made, making it a very rare title and a prize for collectors.

[202] Mott 2013, p. 361.

[203] "The History of Sega Fighting Games". *GameSpot*. p. 11. Retrieved 2014-12-17.; "The History of Sega Fighting Games". *GameSpot*. p. 12. Retrieved 2014-12-17.

[204] Vore, Bryan (2011-10-12). "*Guardian Heroes*". *Game Informer*. Retrieved 2014-03-17. Even though some aspects of *Guardian Heroes* haven't aged well, the strength of the battle system, branching paths, and characters help this brawler retain its place near the top of the class. cf. Parkin, Simon (2011-10-12). "*Guardian Heroes*". *Eurogamer*. Retrieved 2014-03-28. One of the most satisfying combat games ever conceived.

[205] Wallace, Kimberley (2013-04-16). "*Shin Megami Tensei: Devil Summoner: Soul Hackers*". *Game Informer*. Retrieved 2014-03-17. I still can't believe that a game that came out in 1997 feels so fresh and exciting over 15 years later.

[206] "Top 10 Cult Classics". *1UP.com*. 2005-06-22. Archived from the original on 2011-08-04. Retrieved 2014-04-30.

[207] Hatfield, Daemon (2011-09-13). "*Radiant Silvergun* Review". *IGN*. Retrieved 2014-03-15.

[208] "Time Extend: *Radiant Silvergun*". *Edge*. 2013-01-12. Retrieved 2015-03-05.

[209] Ohbuchi, Yutaka (1998-10-21). "*RE2* for Saturn Canceled". *GameSpot*. Retrieved 2014-11-23.

[210] "Suzuki: 'Yes on *VF3*'". *Next Generation*. 1996-11-22. Retrieved 2014-04-21.

[211] "The History of Sega Fighting Games, Page 18". *GameSpot*. Retrieved 2014-06-06. Sega executives started to quietly state that *Virtua Fighter 3* was not going to come out for the Saturn.

[212] "Game Machine Cross Review: セガサターン". *Weekly Famicom Tsūshin* (in Japanese) (335): 166. May 12–19, 1995.

[213] Lynch, Dennis (1995-06-16). "Saturn Runs Rings Around Its Rivals". *Chicago Tribune*. Retrieved 2015-01-20.

[214] Kim, Albert (1995-06-09). "Sega Saturn". *Entertainment Weekly*. Retrieved 2015-01-20.

[215] "EGM rates the systems of 1996!". *1996 Video Game Buyer's Guide*. December 1996.

[216] "Electronic Gaming Monthly looks at the top systems for this year". *1998 Video Game Buyer's Guide*: 51. December 1998.

[217] "How Consoles Die". *Edge*. 2008-09-17. p. 3. Retrieved 2015-03-05.

[218] Stuart, Keith (2015-05-15). "Sega Saturn – how to buy one and what to play". *The Guardian*. Retrieved 2015-05-27.

[219] Towell, Justin (2014-11-22). "Sega Saturn turns 20, and it's not as shit as you think". *GamesRadar*. Retrieved 2015-01-20.

[220] Perry, Douglass C. (2009-09-09). "The Rise and Fall of Dreamcast". *Gamasutra*. Retrieved 2014-12-24.

[221] Buchanan, Levi (2008-07-29). "Top 10 SEGA Saturn Games". *IGN*. Retrieved 2014-03-26.

[222] Stuart, Keith (2015-05-14). "Sega Saturn: how one decision destroyed PlayStation's greatest rival". *The Guardian*. Retrieved 2015-05-27.

### 1.1.6 Bibliography

- Mott, Tony (2013). *1001 Video Games You Must Play Before You Die*. New York, New York: Universe Publishing. ISBN 978-0-7893-2090-2.

- Harris, Blake J. (2014). *Console Wars: Sega, Nintendo, and the Battle That Defined a Generation*. New York, New York: HarperCollins. ISBN 978-0-06-227669-8.

- DeMaria, Rusel; Wilson, Johnny L. (2004). *High Score!: The Illustrated History of Electronic Games*. Emeryville, California: McGraw-Hill/Osborne. ISBN 0-07-223172-6.

- Kent, Steven L. (2001). *The Ultimate History of Video Games: The Story Behind the Craze that Touched our Lives and Changed the World*. Roseville, California: Prima Publishing. ISBN 0-7615-3643-4.

## 1.2 Sega

**Sega Holdings Co., Ltd.** (株式会社セガホールディングス *Kabushiki-gaisha Sega hōrudingusu*), originally short for **Service Games** and officially styled as **SEGA**, is a Japanese multinational video game developer and publisher headquartered in Tokyo, Japan, with multiple offices around the world. Sega developed and manufactured numerous home video game consoles from 1983 to 2001, but the financial losses incurred from their Dreamcast console caused the company to restructure itself in 2001, and focus on providing software as a third-party developer from then on. Nonetheless, Sega remains the world's most prolific arcade producer, with over 500 games in over 70 franchises on more than 20 different arcade system boards since 1981.[*][3]

Sega, along with their sub-studios, are known for their multi-million selling game franchises including *Sonic the Hedgehog*, *Virtua Fighter*, *Phantasy Star*, *Yakuza*, and *Total War*, amongst others. Sega's head offices are located in Ōta, Tokyo, Japan. Sega's North American division, **Sega of America**, is headquartered in Southern California, having moved there from San Francisco, California in 2015. Sega's European division, **Sega Europe**, is headquartered in the Brentford area of London, England.

### 1.2.1 History

See also: List of Sega video game consoles

**Company origins (1940–1982)**

*SEGA Diamond 3 Star*

In 1940, American businessmen Martin Bromley, Irving Bromberg, and James Humpert formed a company called *Standard Games* in Honolulu, Hawaii, to provide coin-operated amusement machines; mostly slot machines to military bases located which they saw as a potential market since due to the onset of World War II, the number of men stationed at the military bases had increased and they would have needed something to pass their spare time. After the war, the company changed its name to *Service Games* due to military focus and seeing Japan which was under Allied occupation as a potential market, started exporting slot machines there to the U.S. military bases. In 1951, when the government of United States started outlawing slot machines, the company moved its base to Tokyo, Japan. There the company provided coin-operated slot machines to U.S. bases in Japan and changed its name again to *Service Games of Japan* in 1952. Soon, the company also

started providing the slot machines for the Japanese public and the company's focus shifted from the U.S. military bases to the Japanese public.*[4]*[5]*[6]*[7]

David Rosen, an American officer in the United States Air Force stationed in Japan, launched a two-minute photo booth business in Tokyo in 1954.*[4] This company eventually became Rosen Enterprises, and in 1957, began importing coin-operated games to Japan. By 1965, Rosen Enterprises grew to a chain of over 200 arcades, with Service Games its only competitor. Rosen then orchestrated a merger between Rosen Enterprises and Service Games, who by then had their own factory facilities, becoming chief executive of the new company, Sega Enterprises, which derived its name from Service Games.*[8]

Within a year, Sega began the transition from importer to manufacturer, with the release of the Rosen designed submarine simulator game, *Periscope*. The game sported light and sound effects considered innovative for that time, eventually becoming quite successful in Japan. It was soon exported to both Europe and the United States, becoming the first arcade game in the US to cost 25 cents per play.*[8]

In 1969, Rosen sold Sega to American conglomerate Gulf and Western Industries, although he remained as CEO following the sale. Under Rosen's leadership, Sega continued to grow and prosper, and in 1972, Gulf and Western made Sega Enterprises a subsidiary, taking the company's stock public. Sega prospered heavily from the arcade gaming boom of the late 1970s, with revenues climbing to over US$100 million by 1979.*[8]

**Entry into the home console market (1982–1989)**

In 1982, Sega's revenues would surpass $214 million, and they introduced the industry's first three-dimensional game, *SubRoc 3D*. The following year, an overabundance of arcade games led to the video game crash, causing Sega's revenues to drop to $136 million. Sega then pioneered the use of laser disks in the video game *Astronbelt*, and designed and released its first home video game console, the SG-1000 for the third generation of home consoles. Despite this, G&W sold the U.S. assets of Sega Enterprises that same year to pinball manufacturer Bally Manufacturing, and in January 1984, Rosen resigned his post with the company.*[8]

The Japanese assets of Sega were purchased for $38 million by a group of investors led by Rosen, Robert Deith, and Hayao Nakayama, a Japanese businessman who owned Esco Boueki (Esco Trading) an arcade game distribution company that had been acquired by Rosen in 1979.*[8]*[9] Nakayama became the new CEO of Sega, Robert Deith Chairman of the Board, and Rosen became head of its subsidiary in the United States. In 1984, the multibillion-dollar

Japanese conglomerate CSK bought Sega, renamed it to Sega Enterprises, headquartered it in Japan, and two years later, shares of its stock were being traded on the Tokyo Stock Exchange. David Rosen's friend, Isao Okawa, the chairman of CSK, became chairman of Sega.[*][8]

Sega would also release the Sega Master System and the first game featuring Alex Kidd, who would be Sega's unofficial mascot until he was replaced by Sonic the Hedgehog in 1991. While the Master System was technically superior to the NES,[*][10] it failed to capture market share in North America and Japan due to highly aggressive strategies by Nintendo and ineffective marketing by Tonka, who marketed the console on behalf of SEGA in the United States.[*][11] However, the Master System was highly successful in Europe, Australia, New Zealand, and Brazil with games still being sold well into the 1990s alongside the Mega Drive and Nintendo's NES and SNES.

In the mid 1980s, Sega released *Hang-On* and *After Burner*, titles that make use of hydraulic cabinet functionality and force feedback control. Sega also released the 360 degree rotating machine R-360. For arcade system boards, Sega released the System series and the Super Scaler series. UFO Catcher was introduced in 1985 and is Japan's most commonly installed claw crane game.[*][12] Sega was also one of the first to introduce medal games with *World Bingo* and *World Derby* in the 1980s, a sub-industry within Japanese arcades up to its current day.

*Sonic the Hedgehog has been Sega's mascot since the character's introduction in 1991.*

### Expansion and mainstream success (1989–2001)

With the introduction of the Sega Genesis in North America in 1989, Sega of America launched an anti-Nintendo campaign to carry the momentum to the new generation of games, with its slogan "Genesis does what Nintendon't." This was initially implemented by Sega of America President Michael Katz.[*][13] When Nintendo launched its Super Nintendo Entertainment System in North America in August 1991, Sega changed its slogan to "Welcome to the next level."

The same year, Sega of America's leadership passed from Michael Katz to Tom Kalinske, who further escalated the "console war" that was developing.[*][14] As a preemptive strike against the release of the SNES, Sega re-branded itself with a new game and mascot, Sonic the Hedgehog. This shift led to a wider success for the Genesis and would eventually propel Sega to 65% of the market in North America for a brief time. Simultaneously, after much previous delay, Sega released the moderately successful Sega CD in Japan in 1991 and in North America in 1992 as a hardware add-on to the Genesis, giving developers the ability to make longer, more sophisticated games. *Sonic the Hedgehog 2* was also released in 1992 for the Genesis, and became the most suc-

cessful game Sega ever produced, selling over six million copies in total.[*][15] During this period, local North American development also increased with the establishments of Sega Technical Institute in 1990, Sega Midwest Studio in 1992, Sega Multimedia Studio in 1993, and the acquisition of Interactive Designs in 1992.

In 1990, Sega launched the Game Gear to compete against Nintendo's Game Boy. However, due to issues with its short battery life, lack of original titles, and weak support from Sega, the Game Gear was unable to surpass the Game Boy, selling approximately 11 million units. The Game Gear was succeeded by the Sega Nomad in 1995 and discontinued in 1997.

In 1992, Sega introduced the Model series of arcade hardware, which saw the release of *Virtua Fighter* and *Virtua Racing*, which laid the foundation for 3D racing and fighting games.[*][16] In 1994, Sega released the Sega 32X in an attempt to upgrade the Genesis to the standards of more advanced systems at the time. It sold well initially, but had problems with lack of software and hype about the upcoming Sega Saturn and Sony's PlayStation.[*][17] Within a year, it was in the bargain bins of many stores.[*][18]

Also in 1994, Sega Channel, a subscription gaming service delivered by local cable companies affiliated with Time Warner Cable, was launched in the United States, through which subscribers received a special cartridge adapter that connected to the cable connection. At its peak, the Sega Channel had approximately 250,000 subscribers. Various technical issues began disrupting the service in late 1997, eventually leading to the Sega Channel being discontinued worldwide in 1998.[*][19]

On November 22, 1994, Sega launched the Sega Saturn in Japan. It utilized two 32-bit processors and preceded both the Sony PlayStation and the Nintendo 64. However, poor sales in the West led to the console being abandoned by 1998.[*][20] The lack of strong titles based on established Genesis franchises, along with its high price in comparison to the Sony PlayStation, were among the reasons for the console's failure.[*][21] Notable titles in Japan include *Radiant Silvergun*, *Sakura Wars*, *Panzer Dragoon*, and arcade ports such as *The House of the Dead*, *Virtua Fighter 2* and *Sega Rally Championship*. *Tomb Raider* was initially developed with the Sega Saturn in mind, but with the Saturn's failure to attract the greater market share, development for the game and its sequels were focused on the PlayStation. With the failure of the Saturn in the West, Sega made forays in the PC market with the establishment of SegaSoft in 1995, which was tasked in creating original PC titles.

In 1996, Sega opened Joypolis, an amusement park and arcade center in Yokohama, Japan, with the overseas versions called SegaWorld. In 1997, Sega entered into a short-lived merger with Bandai. However, it was later called off, citing "cultural differences" between the two companies.[*][22] The first Purikara machine in Japan was made jointly by Atlus and Sega in 1995.

On November 27, 1998, Sega launched the Dreamcast in Japan. The console was competitively priced, partly due to the use of off-the-shelf components, but it also featured technology that allowed for more technically impressive games than its direct competitors, the Nintendo 64 and PlayStation. An analog 56k modem was also included, allowing for online multiplayer. It featured titles such as the action-puzzle title *ChuChu Rocket!*, *Phantasy Star Online*, the first console-based massively multiplayer online roleplaying game (MMORPG), *Quake III Arena* and *Alien Front Online*, the first console game with online voice chat. The Dreamcast's launch in Japan was a failure; launching with a small library of software and in the shadow of the upcoming PS2, the system would gain little ground, despite several successful games in the region.

After closures of all their former American developers in 1995, and the closure of the PC SegaSoft division, Sega invested in the American Visual Concepts and the French No Cliché, although the latter was closed in 2001.

The Dreamcast's Western launch in 1999 was accompanied by a large amount of both first-party and third-party software and an aggressive marketing campaign. In contrast to the Japanese launch, the Western launch earned the distinction of the "most successful hardware launch in history," selling a then-unprecedented 500,000 consoles in its first week in North America.[*][20] Sega was able to hold onto this momentum in the US almost until the launch of Sony's PlayStation 2. The Dreamcast is home to several innovative and critically acclaimed games of the time, including one of the first cel-shaded titles, *Jet Set Radio* (*Jet Grind Radio* in North America); *Seaman*, a game involving communication with a fish-type creature via microphone; *Samba de Amigo*, a rhythm game involving the use of maracas, and *Shenmue*, a large-scope adventure game with freeform gameplay and a detailed in-game city. Sega also produced the NAOMI series, which were the last arcade boards built uniquely rather than being based on existing consoles and PC architecture.

In late 1999, Sega Enterprises Chairman Isao Okawa spoke at an Okawa Foundation meeting, saying that Sega's focus in the future would shift from hardware to software, but adding that they were still fully behind the Dreamcast. On November 1, 2000, Sega changed its company name from *Sega Enterprises* to *Sega Corporation*.[*][23]

### Shift to third-party software development (2001–2005)

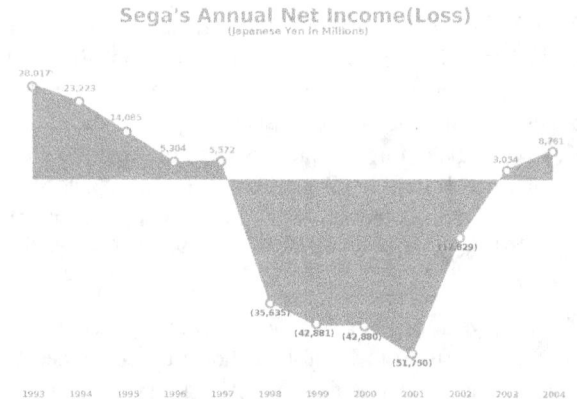

*Sega's financial trouble in the 1998–2002 time periods.*[*][24][*][25][*][26][*][27]

On January 23, 2001, a story ran in *Nihon Keizai Shimbun* claiming that Sega would cease production of the Dreamcast and develop software for other platforms in the future.[*][28] After initial denial, Sega Japan then put out a press release confirming they were considering producing software for the PlayStation 2 and Game Boy Advance as part of their "New Management Policy" .*[29] Subsequently on January 31, 2001, Sega of America officially announced they were becoming a third-party software publisher.*[30] The company has since developed into a third-party publisher that oversees games that launch on game consoles produced by other companies, many of their former rivals, the first of which was a port of *ChuChu Rocket!* to Nintendo's Game Boy Advance. On March 31, 2001, the Dreamcast was discontinued.

By March 31, 2002, Sega had five consecutive fiscal years of net losses.*[31] To help with Sega's debt, CSK founder Isao Okawa, before his death in 2001, gave the company a $692 million private donation,*[32] and talked to Microsoft about a sale or merger with their Xbox division, but those talks failed.*[33] Discussions also took place with Namco, Bandai, Electronic Arts and again with Microsoft. In August 2003, Sammy, one of the biggest pachinko and pachislot manufacturing companies, bought the outstanding 22% of shares that CSK had,*[34] and Sammy chairman Hajime Satomi became CEO of Sega. In the same year, Hajime Satomi stated that Sega's activity will focus on their profitable arcade business as opposed to their loss-incurring home software development sector.*[35] After the decline of the global arcade industry around the 21st century, Sega introduced several novel concepts tailored to the Japanese market. *Derby Owners Club* was the first large-scale satellite arcade machine with IC cards for data storage. Trading card game machines were introduced, with titles such as *World Club Champion Football* for general audiences and *Mushiking: King of the Beetles* for young children. Sega also introduced internet functionality in arcades with *Virtua Fighter 4* in 2001, and further enhanced it with ALL.Net, introduced in 2004.*[36]

During mid-2004, Sammy bought a controlling share in Sega Corporation at a cost of $1.1 billion, creating the new company Sega Sammy Holdings, an entertainment conglomerate. Since then, Sega and Sammy became subsidiaries of the aforementioned holding company, with both companies operating independently, while the executive departments merged.

**Continued expansion and acquisitions (2005–2013)**

In 2005, Sega sold its major western studio Visual Concepts to Take-Two Interactive,*[37] and purchased UK-based developer Creative Assembly, known for its *Total War* series.*[38] In the same year, the Sega Racing Studio was also formed by former Codemasters employees.*[39] In 2006, Sega Europe purchased Sports Interactive, known for its *Football Manager* series.*[40] Sega of America purchased Secret Level in 2006, which was renamed to Sega Studio San Francisco in 2008. In early 2008, Sega announced

that they would re-establish an Australian presence, as a subsidiary of Sega of Europe, with a development studio branded as Sega Studio Australia. In the same year, Sega launched a subscription based flash website called "Play-SEGA" which played emulated versions of Sega Genesis as well original web-based flash games.*[41] It was subsequently shut down due to low subscription numbers.*[42] In 2013, following THQ's bankruptcy, Sega bought Relic Entertainment, known for its *Company of Heroes* series.*[43] Sega has also collaborated with many western studios such as Bizarre Creations, Backbone Entertainment, Monolith, Sumo Digital, Kuju Entertainment, Obsidian Entertainment and Gearbox Software. In 2008, Sega announced the closure of Sega Racing Studio, although the studio was later acquired by Codemasters.*[39] Closures of Sega Studio San Francisco and Sega Studio Australia followed in 2010 and 2012, respectively.

The *Sonic the Hedgehog* series continued to be internationally recognized, having sold 150 million in total,*[1] although the critical reception of games in the series has been mixed.*[44] In 2007, Sega and Nintendo teamed up using Sega's acquired Olympic Games license, to create the *Mario and Sonic at the Olympic Games* series, which has sold over 20 million in total. In the console and handheld business, Sega of Japan found success with the *Yakuza* and *Hatsune Miku: Project DIVA* series of games, amongst others primarily aimed at the Japanese market. In Japan, Sega distributes titles from smaller Japanese game developers and localizations of western titles.*[45]*[46] In 2013, Index Corporation was purchased by Sega Sammy after going bankrupt.*[47] After the buyout, Sega officially split Index, making Atlus, the video game developer and publisher, a wholly owned subsidiary of Sega.*[48] Atlus is known for its *Megami Tensei* and *Persona* series of role-playing games.

For amusement arcades, Sega's most successful games continued to be based on network and card systems. Games of this type include *Sangokushi Taisen* and *Border Break*. Arcade machine sales incurred higher profits than their console, portable, and PC games on a year-to-year basis until 2014.*[49]

In 2004, the GameWorks chain of arcades became owned by Sega, until the chain was sold off in 2011. In 2009, Sega Republic, an indoor theme park in Dubai, opened to the public. In 2010, Sega began providing the 3D imaging for Hatsune Miku's holographic concerts.*[50] In 2013, in cooperation with BBC Earth, Sega opened the first interactive nature simulation museum, Orbi Yokohama in Yokohama, Japan.*[51]

**Company reshuffling and digital market focus (2013–present)**

Due to the decline of packaged game sales both domestically and outside Japan in the 2010s,[*][52] Sega began layoffs and reduction of their Western businesses, such as Sega shutting down five offices based in Europe and Australia on July 1, 2012.[*][53] This was done in order to focus on the digital game market, such as PC and mobile devices.[*][54][*][55] The amount of SKU gradually shrunk from 84 in 2005 to 32 in 2014. Because of the shrinking arcade business in Japan,[*][56] development personnel would also be relocated to the digital game area.[*][57] Sega gradually reduced its arcade centers from 450 facilities in 2005,[*][58] to around 200 in 2015.[*][59]

In the mobile market, Sega released its first app on the ITunes Store with a version of *Super Monkey Ball* in 2008. Since then, the strategies for Asian and Western markets have become independent. The Western line-up consisted of emulations of games and pay-to-play apps, which were eventually overshadowed by more social and free-to-play games, eventually leading to 19 of the older mobile games being pulled due to quality concerns in May 2015.[*][60][*][61] Beginning in 2012, Sega also began acquiring studios for mobile development, with studios such as Hardlight, Three Rings Design, and Demiurge Studios becoming fully owned subsidiaries.

In the 2010s, Sega established operational firms for each of their businesses, in order to streamline operations. In 2012, Sega established Sega Networks for its mobile games; and although separate at first, it merged with Sega Corporation in 2015. Sega Games is structured as a "Consumer Online Company" promoting cross-play between multiple devices, while Sega Networks focuses on developing games for mobile devices.[*][62] In 2012, Sega Entertainment was established for Sega's amusement facility business, and in 2015, Sega Interactive was established for the arcade game business.[*][63] These new divisions would replace the former *Sega Corporation*, and the new *Sega Holdings* would consolidate all entertainment companies from the Sega Sammy group, which became effective April 1, 2015. The exception to this is Sega Live Creation, formed in 2015, which derived from Sega Entertainment and contains the theme park development of the Joypolis and Orbi theme parks. Sega Live Creations is at the newly formed resort section of Sega Sammy.[*][64]

April 2015 also saw Haruki Satomi, grandson of Hajime Satomi, take office as President and CEO of Sega Holdings.[*][65][*][66] In January 2015, Sega of America announced their relocation from San Francisco to Southern California, which was completed by early summer.[*][67] Due to this, Sega of America did not have their own booth at E3 2015.[*][68]

### 1.2.2   Other products and services

Sega is involved in the merchandising of their own intellectual properties, such as *Sonic the Hedgehog* and *The House of the Dead*, as well as unaffiliated anime and manga franchises such as *Bleach*, *Neon Genesis Evangelion*, and *Initial D*.

In 2003, Sega had plans of broadening its franchises to Hollywood co-operating with John Woo,[*][69] but plans fell through.[*][70] In 2015, Sega, together with advertising agency Hakuhodo, established Stories LLC, in which they are tasked in creating various television shows and films based on forty video games by Sega.[*][71] Currently, Sega owns TMS Entertainment, who had business relationships with Sega dating back to 1992.[*][72] Marza Animation Planet spun off Sega's internal CGI production.

Sega Toys was founded when Yonezawa Toys, Japan's largest post-war toy manufacturer, was merged into Sega in 1994. It was briefly known as Sega-Yonezawa until the Yonezawa name was dropped entirely in April 1998.[*][73] Since the early 2000s Sega Toys has become a mostly separate entity from Sega with its own management structure and goals, with some occasional collaboration between the two; Sega and Sega Toys produce the UFO Catcher prize games jointly, where Sega manufactures the arcade equipment, while Sega Toys produces the prizes . They have created toys for children's franchises such as *Oshare Majo: Love and Berry*, *Mushiking: King of the Beetles*, *Lilpri*, *Bakugan*, Jewelpet, *Dinosaur King* and *Hero Bank*. Products by Sega Toys released in the West include the Homestar and the iDog. Sega Toys also inherited the Sega Pico handheld system and produced software for the console.

### 1.2.3   Company executives

**Sega Holdings**

- Hayao Nakayama: Co-founder, President (1984–1998)

- Shoichiro Irimajiri: President (1998–2000)

- Isao Okawa: President (2000–2001)

- Hideki Sato: President (2001-2003)

- Hisao Oguchi: President (2003–2008)

- Okitane Usui: President and COO (2008–2012)

- Toshihiro Nagoshi: Director and CCO (2012–present)

- Naoya Tsurumi: President and COO (2012–present)

- Hideki Okamura: President and COO (2013–2015)

- Haruki Satomi: President and CEO (2015–present)

**Sega of America**

- David Rosen: Co-founder

- Bruce Lowry: President (1986–1988)[*][74]

- Michael Katz: President (1989–1991)

- Tom Kalinske: President (1991–1996)

- Bernie Stolar: President (1996–1999)

- Peter Moore: President (1999–2003)

- Simon Jeffery: President (2003–2009)

- Mike Hayes: President (2009–2012)

- John Cheng: President and COO (2012–present)

**Sega Europe**

- Robert Deith: Co-founder/chairman (1991–2001)

- Naoya Tsurumi: CEO (2005–2009)[*][75][*][76]

- Mike Hayes: President (2009–2012)

- Jürgen Post: President (2012–present)

## 1.2.4 See also

- List of Sega video game consoles

- List of Sega video game franchises

- Lists of Sega games

- Sega Seal of Quality

- Sega Enterprises Ltd. v. Accolade, Inc.

- Sega development studios

## 1.2.5 References

[1] "Sega Sammy Holdings – Annual Report 2014" (PDF). *segasammy.jp*. Sega Sammy Holdings. pp. 34, 58, 62, 65. Retrieved May 6, 2015.

[2] "Corporate Profile". *sammy-net.hp*. Sammy Networks Co., Ltd. Retrieved May 7, 2015.

[3] "Most prolific producer of arcade machines". *Guinness World Records*. Jim Pattison Group. Retrieved May 7, 2015.

[4] Fahs, Travis (21 April 2009). "IGN Presents the History of SEGA". *IGN*. j2 Global. Retrieved 29 July 2015.

[5] Plunkett, Luke (4 April 2011). "Meet the four Americans who built Sega". *Kotaku*. Gawker Media. Retrieved 1 August 2015.

[6] "IBM turns 100: other surprisingly ancient technology companies". *The Guardian*. Scott Trust Limited. Retrieved 1 August 2015.

[7] Daniel Sànchez-Crespo Delmau (2004). *Core Techniques and Algorithms in Game Programming*. New Riders. p. 3. ISBN 9780131020092.

[8] "History of Sega of America, Inc.". *fundinguniverse.com*. Retrieved May 7, 2015.

[9] Pollack, Andrew (July 3, 1993). "Sega Takes Aim at Disney's World". *The New York Times*. The New York Times Company. Retrieved May 7, 2015.

[10] "Sega Master System (SMS) – 1986–1989". Classicgaming.gamespy.com. Archived from the original on 2010-10-17. Retrieved February 23, 2011.

[11] Williams, Mike (November 21, 2013). "Next Gen Graphics, Part 1: NES, Master System, Genesis, and Super NES". *USgamer*. Gamer Network. Retrieved May 7, 2015.

[12] "Sega Sammy Holdings – Annual Report 2005" (PDF). *segasammy.jp*. Sega Sammy Holdings. p. 20. Retrieved May 6, 2015.

[13] Horowitz, Ken (April 28, 2006). "Interview: Michael Katz". *Sega-16.com*. Retrieved May 7, 2015.

[14] Horowitz, Ken (February 18, 2005). "Tom Kalinske: American Samurai". *Sega-16.com*. Retrieved May 7, 2015.

[15] Boutros, Daniel (August 4, 2006). "A Detailed Cross-Examination of Yesterday and Today's Best-Selling Platform Games". *Gamasutra*. UBM plc. p. 5. Retrieved May 7, 2015.

[16] "15 most influential games of all time". *GameSpot*. CBS Interactive. p. 13. Archived from the original on April 12, 2010.

[17] McFerran, Damien (February 22, 2012). "The Rise and Fall of Sega Enterprises". *Eurogamer*. Gamer Network. Retrieved May 7, 2015.

[18] "About – Sega History". *PlanetDreamcast.com*. June 16, 2008. Archived from the original on June 16, 2008. Retrieved February 23, 2011.

[19] Buchanan, Levi (June 11, 2008). "The SEGA Channel". *IGN*. Ziff Davis. Retrieved February 23, 2011.

[20] Whitehead, Dan (January 2, 2009). "Dreamcast: A Forensic Retrospective". *Eurogamer*. Gamer Network. Retrieved May 7, 2015.

[21] Buchanan, Levi (February 2, 2009). "What Hath Sonic Wrought? Vol. 10". *IGN*. Ziff Davis. Retrieved May 7, 2015.

[22] Johnston, Chris (May 27, 1997). "Sega, Bandai Merger Canceled". *GameSpot*. CBS Interactive. Retrieved May 7, 2015.

[23] "Sega Enterprises, Ltd. Changes Company Name". *Sega.jp*. Sega. November 1, 2001. Retrieved May 7, 2015.

[24] "Sega Enterprises, Ltd. Annual Report 1998" (PDF). *Sega.jp*. Sega. p. 8. Archived from the original (PDF) on June 17, 2002.

[25] "Sega Enterprises, Ltd. Annual Report 2000" (PDF). *Sega.jp*. Sega. Archived from the original (PDF) on September 25, 2007. Retrieved March 12, 2010.

[26] "Sega Corporation Annual Report 2002" (PDF). *segasammy.jp*. Sega Sammy Holdings. Retrieved March 12, 2010.

[27] "Sega Corporation Annual Report 2004". *segasammy.jp*. Sega Sammy Holdings. pp. 2, 16. Archived from the original (PDF) on December 25, 2009. Retrieved March 12, 2010.

[28] Justice, Brandon (January 23, 2001). "Sega Sinks Console Efforts?". *IGN*. Ziff Davis. Retrieved May 7, 2015.

[29] Gantayat, Anoop (January 23, 2001). "Sega Confirms PS2 and Game Boy Advance Negotiations". *IGN*. Ziff Davis. Retrieved May 7, 2007.

[30] Ahmed, Shahed (January 31, 2001). "Sega announces drastic restructuring". *GameSpot*. CBS Interactive. Retrieved September 20, 2009.

[31] "Analysts say Sega taking its toll on CSK's bottom line". *Taipei Times*. The Liberty Times Group. March 13, 2003. Retrieved May 7, 2015.

[32] Tanikawa, Miki (March 17, 2001). "Isao Okawa, 74, Chief of Sega And Pioneer Investor in Japan". *The New York Times*. The New York Times Company. Retrieved May 7, 2015.

[33] Gaither, Chris (November 1, 2001). "Microsoft Explores A New Territory: Fun". *The New York Times*. The New York Times Company. p. 2. Retrieved May 7, 2015.

[34] Niizumi, Hirohiko; Thorsen, Tor (May 18, 2004). "Sammy merging with Sega". *GameSpot*. CBS Interactive. Archived from the original on October 6, 2008. Retrieved February 18, 2011.

[35] Bramwell, Tom (December 11, 2003). "Sammy tells Sega to focus on arcade". *Eurogamer*. Gamer Network. Retrieved May 7, 2015.

[36] "Sega Sammy Holdings – Annual Report 2007" (PDF). *segasammy.jp*. Sega Sammy Holdings. p. 36. Retrieved May 7, 2015.

[37] Feldman, Curt; Thorsen, Tor (January 24, 2005). "Sega officially out of the sports game". *GameSpot*. CBS Interactive. Retrieved May 7, 2015.

[38] Bramwell, Tom (March 9, 2005). "SEGA acquires Creative Assembly". *Eurogamer*. Gamer Network. Retrieved May 7, 2015.

[39] Hayward, Andrew (April 25, 2008). "Codemasters Acquires Sega Racing Studio". *1UP.com*. Ziff Davis. Retrieved May 7, 2015.

[40] "SEGA acquires Sports Interactive". *Eurogamer*. Gamer Network. April 4, 2006. Retrieved May 7, 2015.

[41] "Is PlaySega Worth Your Money? | CINEMABLEND". Retrieved 2015-05-31.

[42] "Sarah May Wellock | LinkedIn". *uk.linkedin.com*. Retrieved 2015-05-31.

[43] Goldfarb, Andrew (January 23, 2013). "THQ Dissolved, Saints Row, Company of Heroes Devs Acquired". *IGN*. Ziff Davis. Retrieved May 7, 2015.

[44] "Sonic - Reviews, Articles, People, Trailers and more at Metacritic - Metacritic". *www.metacritic.com*. Retrieved 2015-06-12.

[45] "セガ製品情報" [Sega product information]. *sega.jp* (in Japanese). Sega. Retrieved May 7, 2015.

[46] "Sega PC Localized Game Official Site". *sega.jp*. Sega. Retrieved May 7, 2015.

[47] MacGregor, Kyle (September 19, 2013). "Atlus 'extremely happy' to join forces with Sega". *Destructoid*. Retrieved May 7, 2015.

[48] Pitcher, Jenna (February 18, 2014). "Sega to rebrand Index as Atlus in April, creates new division". *Polygon*. Vox Media. Retrieved May 7, 2015.

[49] "Sales by segment – Financial Information – Investor Relations". *www.segasammy.co.jp*. Sega Sammy Holdings. Retrieved April 5, 2015.

[50] Verini, James (October 19, 2012). "How Virtual Pop Star Hatsune Mikue Blew Up in Japan". *Wired*. Condé Nast. Retrieved May 7, 2015.

[51] Lanxon, Nate (August 20, 2013). "The Orbi story: BBC and Sega collaborate on experimental natural history theme park". *Wired*. Condé Nast. Retrieved May 7, 2015.

[52] Rose, Mike (May 11, 2012). "Sega focusing on digital shift following decreased 2011 financials". *Gamasutra*. UBM plc. Retrieved May 7, 2015.

[53] Harris, Jake (June 28, 2012). "Sega to close five European, Australian offices". *GameSpot*. CBS Interactive. Retrieved May 7, 2015.

[54] Moscritolo, Angela (March 30, 2012). "Sega Cancelling Games, Planning Layoffs". *PC Magazine*. Ziff Davis. Retrieved April 8, 2015.

[55] Crossley (January 30, 2015). "Sega to Axe 300 Jobs as Focus Turns to PC and Mobile". *Yahoo! Games*. Yahoo! first=Rob. Retrieved April 14, 2015.

[56] "Market Data". *capcom.co.jp*. Capcom. Retrieved April 5, 2015.

[57] "Business Strategies". *segasammy.co.jp*. Sega Sammy Holdings. Retrieved April 5, 2015.

[58] Kohler, Chris (October 2, 2009). "Sega to Close Arcades, Cancel Games, Lay Off Hundreds". *Wired*. Condé Nast. Retrieved May 7, 2015.

[59] "FY Ending March 2015 – 3rd Quarter Results Presentation" (PDF). *segasammy.co.jp*. Sega Sammy Holdings. Retrieved April 14, 2015.

[60] "SEGA Mobile Game Closures". Sega Blog. Retrieved May 9, 2015.

[61] Rao, Chloi (May 8, 2015). "SEGA Removing Games From Mobile Catalog that Fail to Meet Quality Standards". IGN. Retrieved May 10, 2015.

[62] "事業内容 | 株式会社セガゲームス". *sega-games.co.jp*. Retrieved May 15, 2015.

[63] "Notice of Organizational Restructuring within the Group and Change of Names of Some Subsidiaries due to the Restructuring" (PDF). *segasammy.co.jp*. Sega Sammy Holdings. February 12, 2015. Retrieved May 7, 2015.

[64] "Group Overview: SEGA SAMMY Group". *www.segasammy.co.jp*. Retrieved May 17, 2015.

[65] "セガゲームス始動！代表取締役社長 CEO 里見治紀氏に訊く新会社設立の意図と将来像". *www.famitsu.com*. Retrieved September 9, 2015.

[66] "Executive Profile | SEGA SAMMY Group | SEGA SAMMY HOLDINGS". *www.segasammy.co.jp*. Retrieved September 9, 2015.

[67] "SEGA of America Relocates to Southern California". *Yahoo! Finance*. Yahoo!. January 30, 2015. Retrieved April 14, 2015.

[68] Futter, Mike. "Sega Will Not Have Its Own Booth At E3 2015". Game Informer. Retrieved May 10, 2015.

[69] "John Woo-Backed Studio Partners With Sega". Retrieved May 15, 2015.

[70] "Hollywood's Long History of Mostly Failing to Make Video Games". Retrieved May 15, 2015.

[71] "STORIES LLC, STORIES INTERNATIONAL INC". *Stories International*. Retrieved May 15, 2015.

[72] "TMS Entertainment - All The Tropes". *allthetropes.orain.org*. Retrieved May 15, 2015.

[73] "Company Profile - History of SEGA TOYS". *sega-toys.co.jp*.

[74] "Bruce Lowry". LinkedIn. Retrieved February 23, 2011.

[75] "SEGA Integrates SEGA of America and SEGA Europe Management Teams To Drive Growth In Western Markets". *gameindustry.biz*. Gamer Network. January 20, 2005. Retrieved May 20, 2014.

[76] Fletcher, JC (June 18, 2009). "Sega's Naoya Tsurumi promoted to lofty new position". *Joystiq*. AOL. Retrieved May 20, 2014.

### 1.2.6 External links

- Sega Holdings website
- Sega Japan's official website
- Sega of America's official website

# Chapter 2

# Technical Details

## 2.1 CD-ROM

A **CD-ROM** /ˌsiːˌdiːˈrɒm/ is a pre-pressed optical compact disc which contains data. The name is an acronym which stands for "**Compact Disc Read-Only Memory**". Computers can read CD-ROMs, but cannot write to CD-ROMs which are not writable or erasable.

Until the mid-2000s, CD-ROMs were popularly used to distribute software for computers and video game consoles. Some CDs, called enhanced CDs, hold both computer data and audio with the latter capable of being played on a CD player, while data (such as software or digital video) is only usable on a computer (such as ISO 9660 format PC CD-ROMs).

The *Yellow Book* is the technical standard that defines the format of CD-ROMs. One of a set of color-bound books that contain the technical specifications for all CD formats, the *Yellow Book*, created by Sony and Philips in 1988, was the first extension of Compact Disc Digital Audio. It adapted the format to hold any form of data.

### 2.1.1 CD-ROM discs

*A computer drive capable of reading CD-ROM discs*

### Media

CD-ROMs are identical in appearance to audio CDs, and data are stored and retrieved in a very similar manner (only differing from audio CDs in the standards used to store the data). Discs are made from a 1.2 mm thick disc of polycarbonate plastic, with a thin layer of aluminium to make a reflective surface. The most common size of CD-ROM is 120 mm in diameter, though the smaller Mini CD standard with an 80 mm diameter, as well as numerous non-standard sizes and shapes (e.g., business card-sized media) are also available.

Data are stored on the disc as a series of microscopic indentations. A laser is shone onto the reflective surface of the disc to read the pattern of pits and lands ( "pits" , with the gaps between them referred to as "lands" ). Because the depth of the pits is approximately one-quarter to one-sixth of the wavelength of the laser light used to read the disc, the reflected beam's phase is shifted in relation to the incoming beam, causing destructive interference and reducing the reflected beam's intensity. This pattern of changing intensity of the reflected beam is converted into binary data.

### Standard

Several formats are used for data stored on compact discs, known as the Rainbow Books. The *Yellow Book*, published in 1988,[*][2] defines the specifications for CD-ROMs, standardized in 1989 as the ISO/IEC 10149 / ECMA−130 standard. The CD-ROM standard builds on top of the original Red Book CD-DA standard for CD audio. Other standards, such as the *White Book* for Video CDs, further define formats based on the CD-ROM specifications. The *Yellow Book* itself is not freely available, but the standards with the corresponding content can be downloaded for free from ISO[*][1] or ECMA.[*][3]

There are several separate standards that define how to structure data files on a CD-ROM. ISO 9660 defines the standard file system for a CD-ROM. ISO 13490 is an improvement on this standard which adds support for non-sequential write-once and re-writeable discs such as CD-R and CD-RW, as well as multiple sessions. The ISO 13346 standard was designed to address most of the shortcomings of ISO 9660,[4] and a subset of it evolved into the UDF format, which was adopted for DVDs. The bootable CD specification, to make a CD emulate a hard disk or floppy disk, is called El Torito.

**CD-ROM format**

Data stored on CD-ROMs follows the standard CD data encoding techniques described in the *Red Book* specification (originally defined for audio CD only). This includes cross-interleaved Reed–Solomon coding (CIRC), eight-to-fourteen modulation (EFM), and the use of pits and lands for coding the bits into the physical surface of the CD.

The data structures used to group data on a CD-ROM are also derived from the *Red Book*. Like audio CDs (CD-DA), a CD-ROM *sector* contains 2,352 bytes of user data, composed of 98 frames, each consisting of 33-bytes (24 bytes for the user data, 8 bytes for error correction, and 1 byte for the subcode). Unlike audio CDs, the data stored in these sectors corresponds to any type of digital data, not audio samples encoded according to the audio CD specification. In order to structure, address and protect this data, the CD-ROM standard further defines two sector modes, Mode 1 and Mode 2, which describe two different layouts for the data inside a sector. A track (a group of sectors) inside a CD-ROM only contains sectors in the same mode, but if multiple tracks are present in a CD-ROM, each track can have its sectors in a different mode from the rest of the tracks. They can also coexist with audio CD tracks as well, which is the case of mixed mode CDs.

Both Mode 1 and 2 sectors use the first 16 bytes for header information, but differ in the remaining 2,336 bytes due to the use of error correction bytes. Unlike an audio CD, a CD-ROM cannot rely on error concealment by interpolation; a higher reliability of the retrieved data is required. To achieve improved error correction and detection, Mode 1, used mostly for digital data, adds a 32-bit cyclic redundancy check (CRC) code for error detection, and a third layer of Reed–Solomon error correction[5] using a Reed-Solomon Product-like Code (RSPC). Mode 1 therefore contains 288 bytes per sector for error detection and correction, leaving 2,048 bytes per sector available for data. Mode 2, which is more appropriate for image or video data (where perfect reliability may be a little bit less important), contains no additional error detection or correction bytes, having therefore 2,336 available data bytes per sector. Note that both modes, like audio CDs, still benefit from the lower layers of error correction at the frame level.[6]

Before being stored on a disc with the techniques described

above, each CD-ROM sector is scrambled to prevent some problematic patterns from showing up.[3] These scrambled sectors then follow the same encoding process described in the *Red Book* in order to be finally stored on a CD.

The following table shows a comparison of the structure of sectors in CD-DA and CD-ROMs:[3]

The net byte rate of a Mode-1 CD-ROM, based on comparison to CD-DA audio standards, is 44,100 Hz × 16 bits/sample × 2 channels × 2,048 / 2,352 / 8 = 153.6 kB/s = 150 KiB/s. This value, 150 KiB/s, is defined as "1× speed". Therefore, for Mode 1 CD-ROMs, a 1× CD-ROM drive reads 150/2 = 75 consecutive sectors per second.

The playing time of a standard CD is 74 minutes, or 4,440 seconds, contained in 333,000 blocks or sectors. Therefore, the net capacity of a Mode-1 CD-ROM is 682 MB or, equivalently, 650 MiB. For 80 minute CDs, the capacity is 737 MB (703 MiB).

**CD-ROM XA extension**

*CD-ROM XA* is an extension of the *Yellow Book* standard for CD-ROMs that combines compressed audio, video and computer data, allowing all to be accessed simultaneously.[7] It was intended as a bridge between CD-ROM and CD-i (Green Book) and was published by Sony and Philips in 1991.[2] "XA" stands for eXtended Architecture.

CD-ROM XA defines two new sector layouts, called Mode 2 Form 1 and Mode 2 Form 2 (which are different from the original Mode 2). XA Mode 2 Form 1 is similar to the Mode 1 structure described above, and can interleave with XA Mode 2 Form 2 sectors; it is used for data. XA Mode 2 Form 2 has 2,324 bytes of user data, and is similar to the standard Mode 2 but with error detection bytes added (though no error correction). It can interleave with XA Mode 2 Form 1 sectors, and it is used for audio/video data.[6] Video CDs, Super Video CDs, Photo CDs, Enhanced Music CDs and CD-i use these sector modes.[8]

The following table shows a comparison of the structure of sectors in CD-ROM XA modes:

**Disc images**

When a disc image of a CD-ROM is created, this can be done in either "raw" mode (extracting 2,352 bytes per sector, independent of the internal structure), or obtaining only the sector's useful data (2,048/2,336/2,352/2,324 bytes depending on the CD-ROM mode). The file size of a disc image created in raw mode is *always* a multiple of 2,352 bytes (the size of a block).[*][9] Disc image formats that store raw CD-ROM sectors include CCD/IMG, CUE/BIN, and MDS/MDF. The size of a disc image created from the data in the sectors will depend on the type of sectors it is using. For example, if a CD-ROM mode 1 image is created by extracting only each sector's data, its size will be a multiple of 2,048; this is usually the case for ISO disc images.

On a 74-minute CD-R, it is possible to fit larger disc images using raw mode, up to $333,000 \times 2,352 = 783,216,000$ bytes (~747 MiB). This is the upper limit for raw images created on a 74 min or ~650 MiB *Red Book* CD. The 14.8% increase is due to the discarding of error correction data.

### Manufacture

Main article: Compact Disc manufacturing

Pre-pressed CD-ROMs are mass-produced by a process of stamping where a glass master disc is created and used to make "stampers", which are in turn used to manufacture multiple copies of the final disc with the pits already present. Recordable (CD-R) and rewritable (CD-RW) discs are manufactured by a different method, whereby the data are recorded on them by a laser changing the properties of a dye or phase transition material in a process that is often referred to as "burning".

### Capacity

*A CD-ROM can easily store the entirety of a paper encyclopedia's words and images, plus audio & video clips*

CD-ROM capacities are normally expressed with binary prefixes, subtracting the space used for error correction data. A standard 120 mm, *700 MB* CD-ROM can actually hold about 737 MB (703 MiB) of data with error correction (or 847 MB total). In comparison, a single-layer DVD-ROM can hold 4.7 GB of error-protected data, more than 6 CD-ROMs.

### 2.1.2 CD-ROM drives

Further information: Optical disc drive

CD-ROM discs are read using CD-ROM drives. A CD-

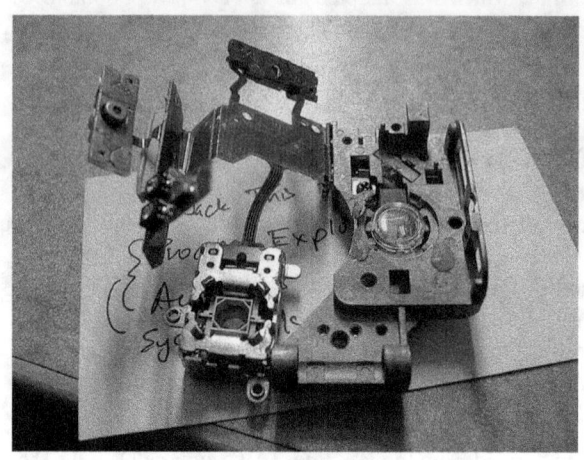

*A view of a CD-ROM drive's disassembled laser system*

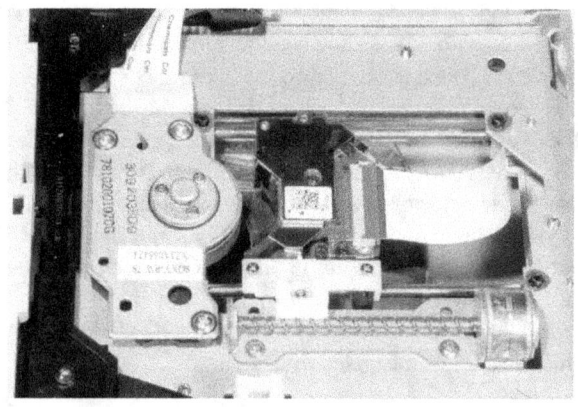

*The movement of the laser enables reading at any position of the CD*

*The laser system of a CD-ROM drive*

ROM drive may be connected to the computer via an IDE (ATA), SCSI, SATA, FireWire, or USB interface or a proprietary interface, such as the Panasonic CD interface. Virtually all modern CD-ROM drives can also play audio CDs (as well as Video CDs and other data standards) when used in conjunction with the right software.

**Laser and optics**

CD-ROM drives employ a near-infrared 780 nm laser diode. The laser beam is directed onto the disc via an optoelectronic tracking module, which then detects whether the beam has been reflected or scattered.

**Transfer rates**

CD-ROM drives are rated with a speed factor relative to music CDs. If a CD-ROM is read at the same rotational speed as an audio CD, the data transfer rate is 150 KiB/s, commonly referred to as "1×". At this data rate, the track moves along under the laser spot at about 1.2 m/s. To maintain this linear velocity as the optical head moves to different positions, the angular velocity is varied from 500 rpm at the inner edge to 200 rpm at the outer edge. The 1× speed rating for CD-ROM (150 KiB/s) is different from the 1× speed rating for DVDs (1.32 MiB/s).

By increasing the speed at which the disc is spun, data can be transferred at greater rates. For example, a CD-ROM drive that can read at 8× speed spins the disc at 1600 to 4000 rpm, giving a linear velocity of 9.6 m/s and a transfer rate of 1200 KiB/s. Above 12× speed most drives read at Constant angular velocity (CAV, constant rpm) so that the motor is not made to change from one speed to another as the head seeks from place to place on the disc. In CAV mode the "×" number denotes the transfer rate at the outer edge of the disc, where it is a maximum. 20× was thought to be the maximum speed due to mechanical constraints until Samsung Electronics introduced the SCR-3230, a 32x CD-ROM drive which uses a ball bearing system to balance the spinning disc in the drive to reduce vibration and noise. As of 2004, the fastest transfer rate commonly available is about 52× or 10,400 rpm and 7.62 MiB/s. Higher spin speeds are limited by the strength of the polycarbonate plastic of which the discs are made. At 52×, the linear velocity of the outermost part of the disk is around 65 m/s. However, improvements can still be obtained by the use of multiple laser pickups as demonstrated by the Kenwood TrueX 72× which uses seven laser beams and a rotation speed of approximately 10×.

Faster 12× drives were common beginning in early 1997. Above 12× speed, there are problems with vibration and heat. CAV drives give speeds up to 30× at the outer edge of the disc with the same rotational speed as a standard constant linear velocity (CLV) 12×, or 32× with a slight increase. However, due to the nature of CAV (linear speed at the inner edge is still only 12×, increasing smoothly inbetween) the actual throughput increase is less than 30/12: in fact, roughly 20× average for a completely full disc, and even less for a partially filled one.

Problems with vibration, owing to limits on achievable symmetry and strength in mass-produced media, mean that CD-ROM drive speeds have not massively increased since the late 1990s. Over 10 years later, commonly available drives vary between 24× (slimline and portable units, 10× spin speed) and 52× (typically CD- and read-only units, 21× spin speed), all using CAV to achieve their claimed "max" speeds, with 32× through 48× most common. Even so, these speeds can cause poor reading (drive error correction having become very sophisticated in response) and even shattering of poorly made or physically damaged media, with small cracks rapidly growing into catastrophic breakages when centripetally stressed at 10,000–13,000 rpm (i.e. 40–52× CAV). High rotational speeds also produce undesirable noise from disc vibration, rushing air and the spindle motor itself. Most 21st-century drives allow forced low speed modes (by use of small utility programs) for the sake of safety, accurate reading or silence, and will automatically fall back if a large number of sequential read errors and retries are encountered.

Other methods of improving read speed were trialled such as using multiple optical beams, increasing throughput up to 72× with a 10× spin speed, but along with other technologies like 90~99 minute recordable media and "double density" recorders, their utility was nullified by the introduction of consumer DVD-ROM drives capable of consistent 36× CD-ROM speeds (4× DVD) or higher. Additionally, with a 700 MB CD-ROM fully readable in under 2½ minutes at 52× CAV, increases in actual data transfer rate are decreasingly influential on overall effective drive speed when taken into consideration with other factors such as loading/unloading, media recognition, spin up/down and random seek times, making for much decreased returns on development investment. A similar stratification effect has since been seen in DVD development where maximum speed has stabilised at 16× CAV (with exceptional cases between 18× and 22×) and capacity at 4.3 and 8.5 GiB (single and dual layer), with higher speed and capacity needs instead being catered to by Blu-ray drives.

CD-Recordable drives are often sold with three different speed ratings, one speed for write-once operations, one for re-write operations, and one for read-only operations. The speeds are typically listed in that order; i.e. a 12×/10×/32× CD drive can, CPU and media permitting, write to CD-R discs at 12× speed (1.76 MiB/s), write to CD-RW discs at 10× speed (1.46 MiB/s), and read from CDs at 32× speed

(4.69 MiB/s).

### 2.1.3  Copyright issues

Main article: Compact Disc and DVD copy protection

Software distributors, and in particular distributors of computer games, often make use of various copy protection schemes to prevent software running from any media besides the original CD-ROMs. This differs somewhat from audio CD protection in that it is usually implemented in both the media and the software itself. The CD-ROM itself may contain "weak" sectors to make copying the disc more difficult, and additional data that may be difficult or impossible to copy to a CD-R or disc image, but which the software checks for each time it is run to ensure an original disc and not an unauthorized copy is present in the computer's CD-ROM drive.

Manufacturers of CD writers (CD-R or CD-RW) are encouraged by the music industry to ensure that every drive they produce has a unique identifier, which will be encoded by the drive on every disc that it records: the RID or Recorder Identification Code.*[11] This is a counterpart to the Source Identification Code (SID), an eight character code beginning with "IFPI" that is usually stamped on discs produced by CD recording plants.

### 2.1.4  See also

- CD/DVD authoring
- Compact Disc Digital Audio
- Computer hardware
- DVD-Audio
- DVD-ROM
- MultiLevel Recording
- Optical disc drive
- Phase-change Dual
- Thor-CD

### 2.1.5  References

[1] ISO (1995). "ISO/IEC 10149:1995 – Information technology – Data interchange on read-only 120 mm optical data disks (CD-ROM)". Retrieved 2010-08-06.

[2] "What is Yellow Book?". Searchstorage.techtarget.com. Retrieved 2013-09-23.

[3] "Data Interchange on Read-only 120 mm Optical Data Disks (CD-ROM)". ECMA. June 1996. Retrieved 2009-04-26.

[4] "Birth Announcement: ISO/IEC 13346 and ISO/IEC 13490". Standards.com. Retrieved 2013-09-23.

[5] Note that the CIRC error correction system used in the CD audio format has two interleaved layers.

[6] McFadden, Andy (2002-12-20). "What is XA? CDPLUS? CD-i? MODE1 vs MODE2? Red/yellow/blue book?". *CD-Recordable FAQ*. Retrieved 2008-05-04.

[7] What are CD-ROM Mode-1, Mode-2 and XA?, Sony Storage Support

[8] "Gateway Support - What is CD-ROM/XA?". Support.gateway.com. Retrieved 2013-09-23.

[9] "Optical Media FAQs" (PDF). Archived from the original (PDF) on 2006-10-22. Retrieved 2007-01-06.

[10] to three significant figures

[11] Schoen, Seth (July 20, 2007). "Harry Potter and the Digital Fingerprints". Electronic Frontier Foundation. Retrieved October 24, 2007.

## 2.2  SuperH

**SuperH** (or **SH**) is a 32-bit reduced instruction set computing (RISC) instruction set architecture (ISA) developed by Hitachi and currently produced by Renesas. It is implemented by microcontrollers and microprocessors for embedded systems.

As of 2015, many of the original patents for the SuperH architecture are expiring and the SH2 CPU has been reimplemented as open source hardware under the name J2.

### 2.2.1  History

The SuperH processor core family was first developed by Hitachi in the early 1990s. Hitachi has developed a complete group of upward compatible instruction set CPU cores. The SH-1 and the SH-2 were used in the Sega Saturn and Sega 32X. These cores have 16-bit instructions for better code density than 32-bit instructions, which was a great benefit at the time, due to the high cost of main memory.

A few years later the SH-3 core was added to the SH CPU family; new features included another interrupt concept, a memory management unit (MMU) and a modified cache concept. The SH-3 core also got a DSP extension, then called SH-3-DSP. With extended data paths for efficient DSP processing, special accumulators and a dedicated MAC-type DSP engine, this core was unifying the DSP and

*SH-2 on Sega 32X and Sega Saturn*

the RISC processor world. A derivative was also used with the original SH-2 core.

Between 1994 and 1996, 35.1 million SuperH devices were shipped worldwide.*[1]

For the Dreamcast, Hitachi developed the SH-4 architecture. Superscalar (2-way) instruction execution and a vector floating point unit were the highlights of this architecture. SH-4 based standard chips were introduced around 1998.

The SH-3 and SH-4 architectures support both big-endian and little-endian byte ordering (they are bi-endian).

**Licensing**

In early 2001, Hitachi and STMicroelectronics formed the IP company SuperH, Inc., which was going to license the SH-4 core to other companies and was developing the SH-5 architecture, the first move of SuperH into the 64-bit area. SuperH, Inc. sold the IP of these CPU cores to Renesas Technology in 2004, which became Renesas Electronics in 2010.

The SH-5 design supported two modes of operation. SHcompact mode is equivalent to the user-mode instructions of the SH-4 instruction set. SHmedia mode is very different, using 32-bit instructions with sixty-four 64-bit integer registers and SIMD instructions. In SHmedia mode the destination of a branch (jump) is loaded into a branch register separately from the actual branch instruction. This allows the processor to prefetch instructions for a branch without having to snoop the instruction stream. The combination of a compact 16-bit instruction encoding with a more powerful 32-bit instruction encoding is not unique to

SH-5; ARM processors have a 16-bit Thumb mode (ARM licensed several patents from SuperH for Thumb*[2]) and MIPS processors have a MIPS-16 mode. However, SH-5 differs because its backward compatibility mode is the 16-bit encoding rather than the 32-bit encoding.

The evolution of the SuperH architecture still continues. The latest evolutionary step happened around 2003 where the cores from SH-2 up to SH-4 were getting unified into a superscalar SH-X core which forms a kind of instruction set superset of the previous architectures.

Today, the SuperH CPU cores, architecture and products are with Renesas Electronics, a merger of the Hitachi and Mitsubishi semiconductor groups and the architecture is consolidated around the SH-2, SH-2A, SH-3, SH-4 and SH-4A platforms giving a scalable family.

**J Core**

The last of the SH-2 patents expired in 2014. At LinuxCon Japan 2015, the Open Processor Foundation presented its cleanroom reimplemention of the SH-2 architecture (known as the "J2 core" due to the unexpired trademarks).*[2]*[3]

The open source VHDL code for the J2 core has been proven on Xilinx FPGAs and on ASICs manufactured on TSMC's 180 nm process, and is capable of booting uClinux.*[2] The Open Processor Foundation plans to implement the SH2A (as "J2+") and SH4 (as "J4") instruction sets as the relevant patents expire in 2016-2017.*[2]

Several features of SuperH have been cited as motivations for designing new cores based on this architecture:*[2]

- High code density compared to other 32-bit RISC ISAs such as ARM or MIPS*[4]

- Existing compiler and operating system support (Linux, Windows Embedded, QNX*[3])

- Low ASIC fabrication costs now that the patents are expiring (around US$0.03 for a J2 core on TSMC's 180 nm process).

### 2.2.2 Models

The family of SuperH CPU cores includes:

- SH-1 - used in microcontrollers for deeply embedded applications (CD-ROM drives, major appliances, etc.)

- SH-2 - used in microcontrollers with higher performance requirements, also used in automotive such as engine control units or in networking applications, and

*Renesas SH-3 CPU*

also in video game consoles, like the Sega Saturn. The SH-2 has also found home in many motor control applications, including Subaru, Mitsubishi, and Mazda.

- SH-2A - The SH-2A core is an extension of the SH-2 core including a few extra instructions but most importantly moving to a superscalar architecture (it is capable of executing more than one instruction in a single cycle) and two five-stage pipelines. It also incorporates 15 register banks to facilitate an interrupt latency of 6 clock cycles. It is also strong in motor control application but also in multimedia, car audio, powertrain, automotive body control and office + building automation

- SH-DSP - initially developed for the mobile phone market, used later in many consumer applications requiring DSP performance for JPEG compression etc.

- SH-3 - used for mobile and handheld applications such as the Jornada, strong in Windows CE applications and market for many years in the car navigation market

- SH-3-DSP - used mainly in multimedia terminals and networking applications, also in printers and fax machines

- SH-4 - used whenever high performance is required such as car multimedia terminals, video game consoles, or set-top boxes

- SH-5 - used in high-end 64-bit multimedia applications

- SH-X - mainstream core used in various flavours (with/without DSP or FPU unit) in engine control unit, car multimedia equipment, set-top boxes or mobile phones

- SH-Mobile - SuperH Mobile Application Processor; designed to offload application processing from the baseband LSI

## SH-2

*Renesas SH-2 CPU*

The SH-2 is a 32-bit RISC architecture with a 16-bit fixed instruction length for high code density and features a hardware multiply–accumulate (MAC) block for DSP algorithms and has a five-stage pipeline.

The SH-2 has a cache on all ROM-less devices.

It provides 16 general purpose registers, a vector-base-register, global-base-register, and a procedure register.

Today the SH-2 family stretches from 32 KB of on-board flash up to ROM-less devices. It is used in a variety of different devices with differing peripherals such as CAN, Ethernet, motor-control timer unit, fast ADC and others.

## SH-2A

The SH-2A is an upgrade to the SH-2 core. It was announced in early 2006.

At launch in 2007 the SH-2A based SH7211 was the world's fastest embedded flash microcontroller running at 160 MHz. It has later been superseded by several newer SuperH devices running at up to 200 MHz.

New features on the SH-2A core include:

- Superscalar architecture: execution of 2 instructions simultaneously

- Harvard architecture

- Two 5-stage pipelines

- 15 register banks for interrupt response in 6 cycles.

- Optional FPU

The SH-2A family today spans a wide memory field from 16 KB up to and includes many ROM-less variations. The devices feature standard peripherals such as CAN, Ethernet, USB and more as well as more application specific peripherals such as motor control timers, TFT controllers and peripherals dedicated to automotive powertrain applications.

**SH-4**

*Renesas SH-4 CPU*

The SH-4 is a 32-bit RISC CPU and was developed for primary use in multimedia applications, such as Sega's Dreamcast and NAOMI game systems. It includes a much more powerful floating point unit and additional built-in functions, along with the standard 32-bit integer processing and 16-bit instruction size.

SH-4 features include:

- FPU with four floating point multipliers, supporting 32-bit single precision and 64-bit double precision floats

- 128-bit floating point bus allowing 3.2 GB/sec transfer rate from the data cache

- 64-bit external data bus with 32-bit memory addressing, allowing a maximum of 4 GB addressable memory with a transfer rate of 800 MB/sec

- Built-in interrupt, DMA, and power management controllers

**SH-5**

The SH-5 is a 64-bit RISC CPU.

### 2.2.3  References

[1] http://segatech.com/technical/cpu/tech_sh4.html

[2] Nathan Willis (June 10, 2015). "Resurrecting the SuperH architecture". LWN.net.

[3] "J Cores". Open Processor Foundation. Retrieved July 10, 2015.

[4] V.M. Weaver (17 March 2015). "Exploring the Limits of Code Density (Tech Report with Newest Results)" (PDF).

### 2.2.4  External links

- Renesas SuperH - Products, Tools, Manuals, App.Notes, Information

- SH-4 CPU Core Architecture by Hitachi & STMicroelectronics

- Linux SuperH development list

- DCTP - Hitachi 200 MHz SH-4

- in-progress Debian port for SH4

## 2.3  Motorola 68000

This article is about the CPU. For the computer, see Sharp X68000.

The **Motorola 68000** ("'sixty-eight-thousand'"; also called the **m68k** or **Motorola 68k**, "*sixty-eight-kay*") is a 16/32-bit<ref "16/32">Motorola Literature Distribution, Phonenix, AZ (1992). *Motorola M68000 Family Programmer's Reference Manual* (PDF). [motorola]. pp. 1–1. ISBN 0-13-723289-6.</ref> CISC microprocessor core designed and marketed by Motorola Semiconductor Products Sector (now Freescale Semiconductor). Introduced in 1979 with HMOS technology as the first member of the successful 32-bit m68k family of microprocessors, it is generally software forward compatible with the rest of the line despite being limited to a 16-bit wide external bus.[1] After 35 years in production, the 68000 architecture is still in use.

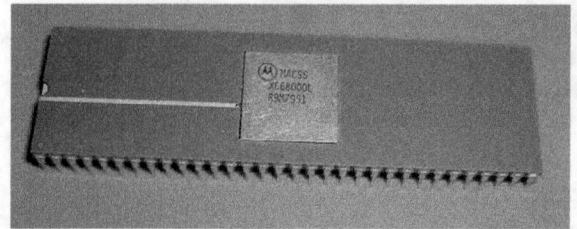

*Pre-release XC68000 chip manufactured in 1979.*

*Die of Motorola 68000.*

*Motorola MC68000 (CLCC package)*

*Motorola MC68000 (PLCC package)*

### 2.3.1  History

The 68000 grew out of the **MACSS** (Motorola Advanced Computer System on Silicon) project, begun in 1976 to develop an entirely new architecture without backward compatibility. It would be a higher-power sibling complementing the existing 8-bit 6800 line rather than a compatible successor. In the end, the 68000 did retain a bus protocol compatibility mode for existing 6800 peripheral devices, and a version with an 8-bit data bus was produced. However, the designers mainly focused on the future, or forward compatibility, which gave the 68000 platform a head start against later 32-bit instruction set architectures. For instance, the CPU registers are 32 bits wide, though few self-contained structures in the processor itself operate on 32 bits at a time. The MACSS team drew heavily on the influence of minicomputer processor design, such as the PDP-11 and VAX systems, which were similarly microcoded.

In the mid 1970s, the 8-bit microprocessor manufacturers raced to introduce the 16-bit generation. National Semiconductor had been first with its IMP-16 and PACE processors in 1973–1975, but these had issues with speed. Intel had worked on their advanced 16/32-bit iAPX432 (alias 8800) since 1975 and their Intel 8086 since 1976 (it was introduced in 1978 but became really widespread in the form of the almost identical 8088 in the IBM PC a few years later). Arriving late to the 16-bit arena afforded the new processor more transistors (roughly 40 000 active versus 20 000 active in the 8086), 32-bit macroinstructions, and acclaimed general ease of use.

The original MC68000 was fabricated using an HMOS

process with a 3.5 μm feature size. Formally introduced in September 1979,[2] Initial samples were released in February 1980, with production chips available over the counter in November.[3] Initial speed grades were 4, 6, and 8 MHz. 10 MHz chips became available during 1981, and 12.5 MHz chips by June 1982.[3] The 16.67 MHz "12F" version of the MC68000, the fastest version of the original HMOS chip, was not produced until the late 1980s. Tom Gunter, retired Corporate Vice President at Motorola, is known as the "Father of the 68000".

The 68k instruction set was particularly well suited to implement Unix,[4] and the 68000 became the dominant CPU for Unix-based workstations including Sun workstations and Apollo/Domain workstations, and also was used for mass-market computers such as the Apple Lisa, Macintosh, Amiga, and Atari ST. The 68000 was used in Microsoft Xenix systems as well as an early NetWare Unix-based Server. The 68000 was used in the first generation of desktop laser printers including the original Apple Inc. LaserWriter and the HP LaserJet. In 1982, the 68000 received an update to its ISA allowing it to support virtual memory and to conform to the Popek and Goldberg virtualization requirements. The updated chip was called the 68010. A further extended version which exposed 31 bits of the address bus was also produced, in small quantities, as the 68012.

To support lower-cost systems and control applications with smaller memory sizes, Motorola introduced the 8-bit compatible MC68008, also in 1982. This was a 68000 with an 8-bit data bus and a smaller (20 bit) address bus. After 1982, Motorola devoted more attention to the 68020 and 88000 projects.

## Second-sourcing

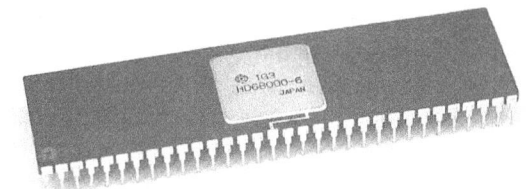

*Hitachi HD68000*

Several other companies were second-source manufacturers of the HMOS 68000. These included Hitachi (HD68000), who shrank the feature size to 2.7 μm for their 12.5 MHz version,[3] Mostek (MK68000), Rockwell (R68000), Signetics (SCN68000), Thomson/SGS-Thomson (originally EF68000 and later TS68000), and

*Thomson TS68000*

Toshiba (TMP68000). Toshiba was also a second-source maker of the CMOS 68HC000 (TMP68HC000).

## CMOS versions

The 68HC000, the first CMOS version of the 68000, was designed by Hitachi and jointly introduced in 1985.[5] Motorola's version was called the MC68HC000, while Hitachi's was the HD68HC000. The 68HC000 was eventually offered at speeds of 8-20 MHz. Except for using CMOS circuitry, it behaved identically to the HMOS MC68000, but the change to CMOS greatly reduced its power consumption. The original HMOS MC68000 consumed around 1.35 watts at an ambient temperature of 25 °C, regardless of clock speed, while the MC68HC000 consumed only 0.13 watts at 8 MHz and 0.38 watts at 20 MHz. (Unlike CMOS circuits, HMOS still draws power when idle, so power consumption varies little with clock rate.) Apple selected the 68HC000 for use in the Macintosh Portable.

Motorola replaced the MC68008 with the MC68HC001 in 1990.[6] This chip resembled the 68HC000 in most respects, but its data bus could operate in either 16-bit or 8-bit mode, depending on the value of an input pin at reset. Thus, like the 68008, it could be used in systems with cheaper 8-bit memories.

The later evolution of the 68000 focused on more modern embedded control applications and on-chip peripherals. The **68EC000** chip and SCM68000 core expanded the address bus to 32 bits, removed the M6800 peripheral bus, and excluded the MOVE from SR instruction from user mode programs.[7] In 1996, Motorola updated the standalone core with fully static circuitry drawing only 2 μW in low-power mode, calling it the MC68SEC000.[8]

Motorola ceased production of the HMOS MC68000 and MC68008 in 1996,[9] but its spin-off company, Freescale Semiconductor, were still producing the MC68HC000, MC68HC001, MC68EC000, and MC68SEC000, as well as the MC68302 and MC68306 microcontrollers and later versions of the DragonBall family. The 68000's architectural descendants, the 680x0, CPU32, and Coldfire families, were also still in production. More recently, with the

Sendai fab closure, all 68HC000, 68020, 68030, and 68882 parts have been discontinued, leaving only the 68SEC000 in production.

### As a microcontroller core

After being succeeded by "true" 32-bit microprocessors, the 68000 was used as the core of many microcontrollers. In 1989, Motorola introduced the MC68302 communications processor.*[10]

## 2.3.2   Applications

*Two Hitachi 68HC000 CPUs being used on an arcade game PCB*

At its introduction, the 68000 was first used in high-priced systems, including multiuser microcomputers like the WICAT 150,*[11] early Alpha Microsystems computers, Sage II / IV, Tandy TRS-80 Model 16, and Fortune 32:16; single-user workstations such as Hewlett-Packard's HP 9000 Series 200 systems, the first Apollo/Domain systems, Sun Microsystems' Sun-1, and the Corvus Concept; and graphics terminals like Digital Equipment Corporation's VAXstation 100 and Silicon Graphics' IRIS 1000 and 1200. Unix systems rapidly moved to the more capable later generations of the 68k line, which remained popular in that market throughout the 1980s.

By the mid-1980s, falling production cost made the 68000 viable for use in personal and home computers, starting with the Apple Lisa and Macintosh, and followed by the Commodore Amiga, Atari ST, and Sharp X68000. On the other hand, the Sinclair QL microcomputer was the most commercially important utilisation of the 68008, along with its derivatives, such as the ICL One Per Desk business terminal. Helix Systems (in Missouri, United States) designed an extension to the SWTPC SS-50 bus, the SS-64, and produced systems built around the 68008 processor.

While the rapid pace of computer advancement quickly rendered the 68000 obsolete as desktop/workstation CPU, the processor found substantial use in embedded applications. By the early 1980s, quantities of 68000 CPUs could be purchased for less than $30 USD per part.

Video game manufacturers used the 68000 as the backbone of many arcade games and home game consoles: Atari's Food Fight, from 1982, was one of the first 68000-based arcade games. Others included Sega's System 16, Capcom's CP System and CPS-2, and SNK's Neo Geo. By the late 1980s, the 68000 was inexpensive enough to power home game consoles, such Sega's Mega Drive (Genesis) console and also the Sega CD attachment for it (A Sega CD system has three CPUs, two of them 68000s). The 1993 multiprocessor Atari Jaguar console used a 68000 as a support chip, although some developers used it as the primary processor due to familiarity. The 1994 multi-processor Sega Saturn console used the 68000 as a sound co-processor (much as the Mega Drive/Genesis uses the Z80 as a co-processor for sound and/or other purposes). Certain arcade games (such as Steel Gunner and others based on Namco System 2) use a dual 68000 CPU configuration, and systems with a triple 68000 CPU configuration also exist (such as Galaxy Force and others based on the Sega Y Board), along with a quad 68000 CPU configuration which has been used by Jaleco (one 68000 for sound has a lower clock rate compared to the others) for titles such as Big Run and Cisco Heat.

The 68000 also saw great success as an embedded controller. As early as 1981, laser printers such as the Imagen Imprint-10 were controlled by external boards equipped with the 68000. The first HP LaserJet—introduced in 1984 —came with a built-in 8 MHz 68000. Other printer manufacturers adopted the 68000, including Apple with its introduction of the LaserWriter in 1985, the first PostScript laser printer. The 68000 continued to be widely used in printers throughout the rest of the 1980s, persisting well into the 1990s in low-end printers.

The 68000 also saw success in the field of industrial control systems. Among the systems which benefited from having a 68000 or derivative as their microprocessor were families of Programmable Logic Controllers (PLCs) manufactured by Allen-Bradley, Texas Instruments and subsequently, following the acquisition of that division of TI, by Siemens. Users of such systems do not accept product obsolescence at the same rate as domestic users and it is entirely likely that despite having been installed over 20 years ago, many 68000-based controllers will continue in reliable service well into the 21st century.

The 683XX microcontrollers, based on the 68000-architecture, are used in networking and telecom equipment, television set-top boxes, laboratory and medical instruments, and even handheld calculators. The MC68302 and its derivatives have been used in many telecom products from Cisco, 3com, Ascend, Marconi, Cyclades and others. Past models of the Palm PDAs and the Handspring Visor used the DragonBall, a derivative of the 68000. AlphaSmart uses the DragonBall family in later versions of its portable word processors. Texas Instruments uses the 68000 in its high-end graphing calculators, the TI-89 and TI-92 series and Voyage 200. Early versions of these used a specialized microcontroller with a static 68EC000 core; later versions use a standard MC68SEC000 processor.

A modified version of the 68000 formed the basis of the IBM XT/370 Hardware emulator of a System 370 processor.

### 2.3.3 Architecture

#### Address bus

The 68000 has a 24-bit external address bus and two byte-select signals "replaced" A0. These 24 lines can therefore reach 16 MB of physical memory with byte resolution. Address storage and computation uses 32 bits internally; however, the 8 high-order address bits are ignored due to the physical lack of device pins. This allows it to run software written for a logically flat 32-bit address space, while accessing only a 24-bit physical address space. Motorola's intent with the internal 32-bit address space was forward compatibility, making it feasible to write 68000 software that would take full advantage of later 32-bit implementations of the 68000 instruction set.[*][1]

However, this did not prevent programmers from writing forward incompatible software. "24-bit" software that discarded the upper address byte, or used it for purposes other than addressing, could fail on 32-bit 68000 implementations. For example, early (pre-7.0) versions of Apple's Mac OS used the high byte of memory-block master pointers to hold flags such as *locked* and *purgeable*. Later versions of the OS moved the flags to a nearby location, and Apple began shipping computers which had "32-bit clean" ROMs beginning with the release of the 1989 Mac IIci.

The 68000 family stores multi-byte binary integers in memory in big-endian order.

#### Internal registers

The CPU has eight 32-bit general-purpose data registers (D0-D7), and eight address registers (A0-A7). The last address register is the stack pointer, and assemblers accept the label SP as equivalent to A7. This was a good number of registers in many ways. It was small enough to allow the 68000 to respond quickly to interrupts (even in the worst case where all 8 data registers D0–D7 and 7 address regis-

ters A0–A6 have to be saved, 15 registers in total), and yet large enough to make most calculations fast, because they can be done entirely within the processor without keeping any partial results in memory. (Note that an exception routine in supervisor mode can also save the user stack pointer A7, which would total 8 address registers. However, the dual stack pointer (A7 and supervisor-mode A7') design of the 68000 makes this normally unnecessary, except when a task switch is performed in a multitasking system.)

Having two types of registers was mildly annoying at times, but not hard to use in practice. Reportedly, it allowed the CPU designers to achieve a higher degree of parallelism, by using an auxiliary execution unit for the address registers.

#### Status register

The 68000 comparison, arithmetic, and logic operations sets bit flags in a status register to record their results for use by later conditional jumps. The bit flags are "zero" (Z), "carry" (C), "overflow" (V), "extend" (X), and "negative" (N). The "extend" (X) flag deserves special mention, because it is separate from the carry flag. This permits the extra bit from arithmetic, logic, and shift operations to be separated from the carry for flow-of-control and linkage.

#### Instruction set

The designers attempted to make the assembly language orthogonal. That is, instructions are divided into operations and address modes, and almost all address modes are available for almost all instructions. There are 56 instructions and a minimum instruction size of 16 bits. Many instructions and addressing modes are longer to include additional address or mode bits.

#### Privilege levels

The CPU, and later the whole family, implements two levels of privilege. User mode gives access to everything except the interrupt level control. Supervisor privilege gives access to everything. An interrupt always becomes supervisory. The supervisor bit is stored in the status register, and is visible to user programs.

An advantage of this system is that the supervisor level has a separate stack pointer. This permits a multitasking system to use very small stacks for tasks, because the designers do not have to allocate the memory required to hold the stack frames of a maximum stack-up of interrupts.

**Interrupts**

The CPU recognizes seven interrupt levels. Levels 1 through 7 are strictly prioritized. That is, a higher-numbered interrupt can always interrupt a lower-numbered interrupt. In the status register, a privileged instruction allows one to set the current minimum interrupt level, blocking lower or equal priority interrupts. For example, if the interrupt level in the status register is set to 3, higher levels from 4 to 7 can caused an exception. Level 7 is a level triggered Non-maskable interrupt (NMI). Level 1 can be interrupted by any higher level. Level 0 means no interrupt. The level is stored in the status register, and is visible to user-level programs.

Hardware interrupts are signaled to the CPU using three inputs that encode the highest pending interrupt priority. A separate Encoder is usually required to encode the interrupts, though for systems that do not require more than three hardware interrupts it is possible to connect the interrupt signals directly to the encoded inputs at the cost of additional software complexity. The interrupt controller can be as simple as a 74LS148 priority encoder, or may be part of a VLSI peripheral chip such as the MC68901 Multi-Function Peripheral (used in the Atari ST range of computers), which also provided a UART, timer, and parallel I/O.

The "exception table" (interrupt vector table interrupt vector addresses) is fixed at addresses 0 through 1023, permitting 256 32-bit vectors. The first vector (RESET) consists of 2 vectors, namely the starting stack address, and the starting code address. Vectors 3 through 15 are used to report various errors: bus error, address error, illegal instruction, zero division, CHK and CHK2 vector, privilege violation (to block privilege escalation), and some reserved vectors that became line 1010 emulator, line 1111 emulator, and hardware breakpoint. Vector 24 starts the **real** interrupts: spurious interrupt (no hardware acknowledgement), and level 1 through level 7 autovectors, then the 16 TRAP vectors, then some more reserved vectors, then the user defined vectors.

Since at a minimum the starting code address vector must always be valid on reset, systems commonly included some nonvolatile memory (e.g. ROM) starting at address zero to contain the vectors and bootstrap code. However, for a general purpose system it is desirable for the operating system to be able to change the vectors at runtime. This was often accomplished by either pointing the vectors in ROM to a jump table in RAM, or through use of bank switching to allow the ROM to be replaced by RAM at runtime.

The 68000 does not meet the Popek and Goldberg virtualization requirements for full processor virtualization because it has a single unprivileged instruction "MOVE from SR", which allows user-mode software read-only access to a small amount of privileged state.

The 68000 is also unable to easily support virtual memory, which requires the ability to trap and recover from a failed memory access. The 68000 does provide a bus error exception which can be used to trap, but it does not save enough processor state to resume the faulted instruction once the operating system has handled the exception. Several companies did succeed in making 68000-based Unix workstations with virtual memory that worked by using two 68000 chips running in parallel on different phased clocks. When the "leading" 68000 encountered a bad memory access, extra hardware would interrupt the "main" 68000 to prevent it from also encountering the bad memory access. This interrupt routine would handle the virtual memory functions and restart the "leading" 68000 in the correct state to continue properly synchronized operation when the "main" 68000 returned from the interrupt.

These problems were fixed in the next major revision of the 68k architecture, with the release of the MC68010. The Bus Error and Address Error exceptions push a large amount of internal state onto the supervisor stack in order to facilitate recovery, and the MOVE from SR instruction was made privileged. A new unprivileged "MOVE from CCR" instruction is provided for use in its place by user mode software; an operating system can trap and emulate user-mode MOVE from SR instructions if desired.

### 2.3.4  Instruction set details

The standard addressing modes are:

- Register direct

  - data register, e.g. "D0"
  - address register, e.g. "A6"

- Register indirect

  - Simple address, e.g. (A0)
  - Address with post-increment, e.g. (A0)+
  - Address with pre-decrement, e.g. -(A0)
  - Address with a 16-bit signed offset, e.g. 16(A0)
  - Register indirect with index register & 8-bit signed offset e.g. 8(A0, D0) or 8(A0, A1)

    Note that with (A0)+ and -(A0), the actual increment or decrement value is dependent on the operand size: a byte access increments the address register by 1, a word by 2, and a long by 4.

- PC (program counter) relative with displacement

- Relative 16-bit signed offset, e.g. 16(PC). This mode was very useful for position-independent code.

- Relative with 8-bit signed offset with index, e.g. 8(PC, D2)

- Absolute memory location

  - Either a number, e.g. "$4000", or a symbolic name translated by the assembler

  - Most 68000 assemblers used the "$" symbol for hexadecimal, instead of "0x" or a trailing H.

  - There were 16 and 32-bit version of this addressing mode

- Immediate mode

  - Data stored in the instruction, e.g. "#400"

- Quick Immediate mode

  - 3 bit unsigned (or 8 bit signed with moveq) with value stored in Opcode

  - In addq and subq, 0 is the equivalent to 8

  - e.g. moveq #0,d0 was quicker than clr.l d0 (though both made d0 equal 0)

Plus: access to the status register, and, in later models, other special registers.

Most instructions have dot-letter suffixes, permitting operations to occur on 8-bit bytes (".b"), 16-bit words (".w"), and 32-bit longs (".l").

Most instructions are **dyadic**, that is, the operation has a source, and a destination, and the destination is changed. Notable instructions were:

- Arithmetic: ADD, SUB, MULU (unsigned multiply), MULS (signed multiply), DIVU, DIVS, NEG (additive negation), and CMP (a sort of comparison done by subtracting the arguments and setting the status bits, but did not store the result)

- Binary Coded Decimal Arithmetic: ABCD, and SBCD

- Logic: EOR (exclusive or), AND, NOT (logical not), OR (inclusive or)

- Shifting: (logical, i.e. right shifts put zero in the most significant bit) LSL, LSR, (arithmetic shifts, i.e. sign-extend the most significant bit) ASR, ASL, (Rotates through eXtend and not:) ROXL, ROXR, ROL, ROR

- Bit test and manipulation in memory: BSET (to 1), BCLR (to 0), BCHG (invert Bit) and BTST (set the Zero bit if tested bit is 0)

- Multiprocessing control: TAS, test-and-set, performed an indivisible bus operation, permitting semaphores to be used to synchronize several processors sharing a single memory

- Flow of control: JMP (jump), JSR (jump to subroutine), BSR (relative address jump to subroutine), RTS (return from subroutine), RTE (return from exception, i.e. an interrupt), TRAP (trigger a software exception similar to software interrupt), CHK (a conditional software exception)

- Branch: Bcc (a branch where the "cc" specified one of 16 tests of the condition codes in the status register: equal, greater than, less-than, carry, and most combinations and logical inversions, available from the status register).

- Decrement-and-branch: DBcc (where "cc" was as for the branch instructions) which decremented a D-register and branched to a destination provided the condition was still true *and* the register had not been decremented to −1. This use of −1 instead of 0 as the terminating value allowed the easy coding of loops which had to do nothing if the count was 0 to begin with, without the need for an additional check before entering the loop. This also facilitated nesting of DBcc.

### 2.3.5 68EC000

The 68EC000 is a low-cost version of the 68000, designed for embedded controller applications. The 68EC000 can have either a 8-bit or 16-bit data bus, switchable at reset.[12]

The processors are available in a variety of speeds including 8 and 16 MHz configurations, producing 2,100 and 4,376 Dhrystones each. These processors have no floating point unit and it is difficult to implement an FPU coprocessor (MC68881/2) with one because the EC series lacks necessary coprocessor instructions.

The 68EC000 was used as a controller in many audio applications, including Ensoniq musical instruments and sound cards where it was part of the MIDI synthesizer.[13] On Ensoniq sound boards, the controller provided several advantages compared to competitors without a CPU on board. The processor allowed the board to be configured to perform various audio tasks, such as MPU-401 MIDI synthesis or MT-32 emulation, without the use of a TSR program.

*Motorola 68EC000 controller*

This improved software compatibility, lowered CPU usage, and eliminated host system memory usage.

The Motorola 68EC000 core was later used in the m68k-based DragonBall processors from Motorola/Freescale.

It also was used as a sound controller in the Sega Saturn game console, and as a controller for the HP JetDirect Ethernet controller boards for the mid-90s LaserJet printers.

### 2.3.6   Example code

The 68000 assembler code below is for a subroutine named strtolower which copies a source null-terminated ASCIZ character string to another destination string, converting all alphabetic characters to lower case.

; strtolower: ; Copy a null-terminated ASCII string, converting ; all alphabetic characters to lower case. ; ; Entry parameters: ; (SP+0): Source string address ; (SP+4): Target string address org $00100000 ;Start at 00100000 00100000 strtolower public 00100000 CE56 0000 link a6,#0 ;Set up stack frame 00100004 206E 0008 movea 8(a6),a0 ;A0 = src, from stack 00100008 226E 000C movea 12(a6),a1 ;A1 = dst, from stack 0010000C 1018 loop move.b (a0)+,d0 ;Load D0 from (src) 0010000E 0C00 0041 cmpi #'A',d0 ;If D0 < 'A', 00100012 650A blo copy ;skip 00100014 0C00 0059 cmpi #'Z',d0 ;If D0 > 'Z', 00100018 6204 bhi copy ;skip 0010001A 0600 0020 addi #'a'-'A',d0 ;D0 = lowercase(D0) 0010001E 12E0 copy move.b d0,(a1)+ ;Store D0 to (dst) 00100020 66E8 bne loop ;Repeat while D0 <> NUL 00100022 4E5E

unlk a6 ;Restore stack frame 00100024 4E75 rts ;Return 00100026 end

The subroutine establishes a call frame using register A6 as the frame pointer. This kind of calling convention supports reentrant and recursive code, and is typically used by languages like C and C++. The subroutine then retrieves the parameters passed to it (src and dst) from the stack. It then loops, reading an ASCII character (a single byte) from the src string, checking if it is an alphabetic character, and if so converting it into a lower-case character, then writing the character into the dst string. Finally, it checks if the character was a null character; if not, it repeats the loop, otherwise it restores the previous stack frame (and A6 register) and returns. Note that the string pointers (registers A0 and A1) are auto-incremented in each iteration of the loop.

In contrast, the code below is for a stand-alone function, even on the most restrictive version of AMS for the TI-89 series of calculators, being kernel-independent, with no values looked up in tables, files or libraries when executing, no system calls, no exception processing, minimal registers to be used, nor the need to save any. It is valid for historical Julian dates from 1 March 1 AD, or for Gregorian ones. In less than two dozen operations it calculates a day-number compatible with ISO 8601 when called with three inputs at their corresponding locations:

; ; WDN, an address - for storing result d0 ; FLAG, 0 or 2 - to choose between Julian or Gregorian, respectively ; DATE, year0mda - date stamp as binary word&byte&byte in basic ISO-format ; YEAR, year (YEAR = DATE due to big-endianess) ; move.l DATE,d0 move.l d0,d1 ; ; APPLY STEP 1 - LACHMAN'S METHOD OF CONGRUENCE andi.l #$f00,d0 divu #100,d0 addi.w #193,d0 andi.l #$ff,d0 divu #100,d0 ; d0 contains the Month Index in the upper word ; ; APPLY STEP 2 - USING SPQR AS A JULIAN YLLD swap d0 andi.l #$ffff,d0 add.b d1,d0 add.w YEAR,d0 subi.l #$300,d1 lsr #2,d1 swap d1 add.w d1,d0 ; SPQR/4 + year + MI + da ; ; (APPLY STEP "0" - GREGORIAN ADJUSTMENT) mulu FLAG,d1 divu #50,d1 mulu #25,d1 lsr #2,d1 add.w d1,d0 add.w FLAG,d0 ; (SP32div16) + SPQR/4 + year + MI + da ; divu #7,d0 swap d0 ; d0.w becomes the day-number ; move.w d0,WDN ; returns the day-number to location WDN rts ; ; Days of the week correspond to day-numbers of the week as: ; Sun=0 Mon=1 Tue=2 Wed=3 Thu=4 Fri=5 Sat=6 ;

### 2.3.7   See also

- Motorola 68000 family

- Motorola 6800, 8-bit predecessor

- Motorola 6809, 8-bit

- Freescale 68HC11, 8-bit embedded

- MacsBug, low-level assembly/machine-level debugger

- x86, the Intel competitor

- Zilog Z8000, 16-bit

- Transistor count

- Instructions per second

### 2.3.8 Notable systems

- The original Apple Macintosh and early successors use the 68000 processor as their CPU.

- The 1981 Sun Workstation and its subsequent commercial spinoff the 1982 Sun-1 workstation used the 68000 as their CPU.

- The Sega Genesis game console uses a 68000 processor (clocked at 7.67 MHz—15/7× the NTSC video colorburst frequency) as its main CPU, and the Sega CD attachment for it uses another 68000 (clocked at 12.5 MHz).

- Neo Geo

- The Commodore Amiga 1000 and 500 use the 68000 processor as their CPU.*[14]

- The Atari ST computers use the 68000 processor, initially with a clock speed of 8 MHz, and later switchable to 16 MHz in the Mega STe.

- CDTV, the world's first compact disc based multimedia platform, uses the 68000 processor as its CPU.*[15]

- The TI-89 Graphing Calculator uses the 68000 processor at 10, 12, or 16 MHz, depending on the calculator's hardware version.

### 2.3.9 References

[1] Starnes, Thomas (April 1983). "Design Philosophy Behind Motorola's MC68000" . *easy68k.com*. BYTE Publications Inc. Retrieved 2015-02-04.

[2] Ken Polsson. "Chronology of Microprocessors" . Processortimeline.info. Retrieved 2013-09-27.

[3] *DTACK GROUNDED, The Journal of Simple 68000/16081 Systems* (29), March 1984, p. 9

[4] Byte Sept 1986

[5] "Company Briefs" , The New York Times, September 21, 1985, available from TimesSelect (subscription).

[6] "68HC001 obsoletes 68008" ., Microprocessor Report, June 20, 1990; available from HighBeam Research (subscription). Archived May 16, 2013 at the Wayback Machine

[7] "Motorola streamlines 68000 family; "EC" versions of 68000, '020, '030, and '040, plus low-end 68300 chip" ., Microprocessor Report, April 17, 1991; available from HighBeam Research (subscription). Archived May 16, 2013 at the Wayback Machine

[8] "Motorola reveals MC68SEC000 processor for low power embedded applications" at the Wayback Machine (archived March 28, 1997), Motorola press release, November 18, 1996.

[9] comp.sys.m68k Usenet posting, May 16, 1995; also see other posts in thread. The end-of-life announcement was in late 1994; according to standard Motorola end-of-life practice, final orders would have been in 1995, with final shipments in 1996.

[10] "Multiprotocol processor marries 68000 and RISC" ., ESD: The Electronic System Design Magazine, November 1, 1989; available from AccessMyLibrary.

[11] "museum ~ WICAT 150" . Old-computers.com. Retrieved 2013-09-27.

[12] Boys, Robert. M68k Frequently Asked Questions (FAQ), comp.sys.m68k, October 19, 1994.

[13] Soundscape Elite Specs. from Fax Sheet, Google Groups, April 25, 1995.

[14] "Commodore Amiga 1000 computer" . Oldcomputers.net. Retrieved 2013-09-27.

[15] Gareth Knight (2002-08-13). "Amiga CDTV" . Amigahistory.co.uk. Retrieved 2013-09-27.

### 2.3.10 Further reading

- *M68000 Microprocessor Users Manual (9th Edition)*; Motorola (Freescale); 224 pages; 1996.

- *Addendum to M68000 User Manual (Rev 0)*; Motorola (Freescale); 26 pages; 1997.

- *M68000 Family Programmer's Reference Manual*; Motorola (Freescale); 646 pages; 1991; ISBN 978-0137232895.

### 2.3.11 External links

- comp.sys.m68k FAQ

- Descriptions of assembler instructions

- 68000 images and descriptions at cpu-collection.de

- 'Chips : Of Diagnostics & Debugging' Article

- EASy68K, an open-source 68k assembler for Windows.

- Feralcore, an open-source 68k emulator, disassembler, and debugger for Java.

- Kiwi - a 68k Homebrew Computer

- m88k - Motorola's VME based 68k boards

## 2.4 Sega NetLink

"NetLink" redirects here. For the Linux program, see Netlink.

**Sega NetLink** (also called **Sega Saturn NetLink**) was an

*The Sega NetLink*

attachment for the Sega Saturn game console to provide Saturn users with internet access and access to email through their console. The unit was released on October 31, 1996. The Sega NetLink consisted of a 28.8 kbit/s modem that fit into the Sega Saturn cartridge port and came packed with a browser developed by Planetweb, Inc. The unit sold for US$199, or US$400 bundled with a Sega Saturn.

The NetLink connected to the internet through standard dial-up services. Unlike other online gaming services in the US, one does not connect to a central service, but instead tells the dial-up modem connected to the Saturn's cartridge slot to call to the person with whom one wishes to play. Since it requires no servers to operate, the service, in theory,

can operate as long as at least two users have the necessary hardware and software, as well as a phone line. [*][1]

In Japan, however, gamers did connect through a centralized service known as SegaNet, which would later be taken offline and converted for Dreamcast usage.

### 2.4.1 Product details

While the NetLink was not the first accessory which allowed console gamers in North America to play video games online, it was the first to allow players to use their own Internet Service Provider (ISP) to connect. While Sega recommended that players use Concentric, the Sega NetLink enabled players to choose any ISP that was within its technical specifications. The device was capable of connecting at a 28.8 kilobit/s connection in America and 14.4 kbit/s in Japan. The success of the Net Link was limited by factors such as high cost, the small number of Saturn owners compared to the competition, and lack of games that took advantage of Net Link capabilities.

Online NetLink games used XBAND technology, which had previously been used in the Super Nintendo Entertainment System and the Sega Genesis modem games.

In Japan, the NetLink required the use of smartcards with prepaid credits. These smartcards or "Saturn media cards" cost ¥2,000 and one game credit was ¥20, which means that one could play about 100 games per card. The Saturn had a floppy drive and printer cable converter (both Japan only) which could be used with the NetLink. A web browser from Planetweb was included, and a mouse and keyboard adapter were available to simplify navigation.

Before the NetLink was not made available for wide release in Europe, Sega performed a test release in Finland before deciding not to widely release the unit in the European market.

Despite the Saturn's relative lack of success in America, the NetLink had a number of users, and five games were released domestically that supported it. Sega of America originally wanted to sell 100,000 NetLinks. [*][2] Sales records show that the Sega sold 40,000 units. [*][3]

In theory, NetLink games can still be played today, as the NetLink modem can use direct-call to connect two players to each other, but the "NetLink Zone" method, which allowed gamers to meet in IRC, can no longer be used as the servers were shut down in 2001.

### 2.4.2 NetLink Zone

The NetLink Zone connected to an Internet Relay Chat server irc.sega.com which was changed to the server irc0.dreamcast.com on the release of Sega's Dreamcast. These servers were originally run by Sega employees but were given over to be run by NetLink chat users Leo Daniels and Mark Leatherman.

### 2.4.3 Successor

SegaNet was launched in 2000 for the Dreamcast, carrying the same name in Japan. The European counterpart was called Dreamarena.

### 2.4.4 Games compatible with NetLink

- Daytona USA CCE NetLink Edition
- Duke Nukem 3D
- Saturn Bomberman
- Sega Rally
- Virtual On

### 2.4.5 See also

- Sega Meganet
- SegaNet

### 2.4.6 Notes

[1] Vinciguerra, Robert. "Discovering the World Through a Sega Saturn NetLink". *The Rev. Rob Times*. Retrieved 16 December 2013.

[2] News article from cnet written before release http://news.cnet.com/Sega-catapults-to-the-Net/ 2100-1023_3-239291.html.

[3] Soohoo, Ken. "Rare Sega Saturn File 4.035 Golden Net Link Web Browser Resurfaces; Plus Republished 1998 Interview With Then Planetweb CTO Ken Soohoo". *The Rev. Rob Times*. Retrieved 16 December 2013.

### 2.4.7 External links

- Planetweb
- Learn Planetweb's Self-Download Feature
- Sega Saturn NetLink League: Information about the NetLink and a resource to find other NetLink players
- Netlink & Dreamcast Old Users @ Way2Live4U.com

# Chapter 3

# Noteworthy Games

## 3.1 Virtua Fighter 2

*Virtua Fighter 2* (Japanese: バーチャファイター2 Hepburn: *Bācha Faitā Tsū*) is a fighting game developed by Sega. It is the sequel to *Virtua Fighter* and the second game in the *Virtua Fighter* series. It was created by Sega's Yu Suzuki-headed AM2 and was released in the arcade in 1994. It was ported to the Sega Saturn in 1995 and Microsoft Windows in 1997. In 1996, a super deformed version of the game, *Virtua Fighter Kids*, arrived in arcades and was ported to the Sega Saturn. A 2D remake was released for the Mega Drive/Genesis in 1996. In addition, *Virtua Fighter 2* was converted for the PlayStation 2 in 2004 as part of Sega's *Ages 2500* series in Japan. The Mega Drive/Genesis port was re-released on the PS2 and PSP in 2006 as part of *Sega Genesis Collection*, on the Virtual Console for the Wii on March 20, 2007 (Japan) and April 16, 2007 (North America), and for iOS on January 20, 2011.

*Virtua Fighter 2* was known for its breakthrough graphics. It used Sega's Model 2 arcade hardware to run the game at 60 frames per second at a high resolution with no slowdown (by comparison, the original *Virtua Fighter* ran at 30 frames per second).*[4] It introduced the use of texture-mapped 3D characters,*[5] and motion capture animation technology.*[6] The Saturn version was also well-received for its graphics and gameplay. It became a huge hit in Japan and sold relatively well in other markets, notably the UK, where *The Prince* (Hatim Habashi) was crowned by Sega Europe as the Official UK *Virtua Fighter 2* Champion.

The arena size could be adjusted up to a very small platform or all the way to 82 meters. This is the only game in the series—other than *Virtua Fighter Remix*—that could have such size adjustments. The physical energy meter could also be adjusted to infinity, giving the player the advantage when beating opponents or practicing moves against the computer player. Adjusting the arena to a smaller size and giving the characters infinite health could lead to mock sumo matches, wherein victory is achieved by knocking the other player's character out of the ring.

### 3.1.1 Characters

Returning characters:

- Akira Yuki
- Pai Chan
- Lau Chan
- Wolf Hawkfield
- Jeffry McWild
- Kage-Maru
- Sarah Bryant
- Jacky Bryant
- Dural

New characters:

- Shun Di
- Lion Rafale

### 3.1.2 Development

At the beginning of 1995, Sega AM2's Sega Saturn division was split into three sub-departments, each one charged with porting a different arcade game to the Saturn: *Virtua Fighter 2*, *Virtua Cop*, and *Daytona USA*. Due to unexpectedly slow progress in the *Daytona USA* port, a number of members of the *Virtua Fighter 2* team were reassigned to *Daytona USA*. In March, AM2 Research completed the Sega Graphics Library, a Saturn operating system which made it feasible to create a near-arcade perfect port of *Virtua Fighter 2* for the Saturn.*[7]*[8]

After completing the *Daytona USA* port in April, the team took a short holiday before beginning work on the *Virtua*

*Fighter 2* conversion in earnest.[*][8] In June, AM2 gave the first public demonstration of Saturn *Virtua Fighter 2* at the Tokyo Toy Show. To increase confidence in the accuracy of the port, they displayed non-playable demos of the characters Lion, Shun, Pai, and Lau running on the Saturn hardware at 60 frames per second - the same speed as the arcade version.[*][7]

However, AM2 continued to face problems in creating an accurate port for the Saturn. Due to the high number of moves in *Virtua Fighter 2*, months had to be spent on developing compression techniques in order to fit all of the game's moves onto a single CD.[*][7] Also, in order to maintain the 60 frames per second, the Saturn version could not use nearly as many polygons as the arcade version. To make this difference less apparent, the programming team used texture mapping on the characters, taking advantage of the fact that the Saturn could map 16 different colors to each polygon, whereas the Model Two arcade hardware could map only 1. In addition, the polygon background objects of the arcade version were replaced with parallax scrolling playfields with selective scaling.[*][7] The AM2 team also used data from *Virtua Fighter Remix* as a reference for some elements.[*][9] In an interview during development, Keiji Okayasu discussed the team's struggles with getting the Saturn version to run at 60 frames per second:

> If we didn't have to consider the speed, we could do the conversion very quickly. But with so much data, we can only move slowly. With *Virtua Fighter 1* we could use the arcade data for each technique with just a few changes, but with 2 there's just too much data. But we have done well, although how is a secret... I think we couldn't have made 2 if we hadn't made the first conversion - but it's just as tough! We owe a lot to the new SGL OS [Sega Graphics Library Operating System] software.[*][10]

By the end of September, hit detection had been enabled, and the now fully playable conversion was displayed at the JAMMA show.[*][8] Taking into account audience reactions at the JAMMA show, the team spent the next two months on final adjustments, play-testing, and the addition of Saturn-specific options. Development on the port was completed in November 1995.[*][8]

### 3.1.3 Release

*Virtua Fighter 2.1* is a revised version featuring re-tweaked gameplay, slightly enhanced graphics and the ability to play as Dural.[*][11] Though it was never released outside of Japan,[*][12] it is possible to switch to the 2.1 game mechanics in the Saturn and PC ports, however none of the other

features are updated. This version was also released in the Sega Ages 2500 series.

In Japan, a Virtua Fighter 2 "CG Portrait Series" of discs were released for the Saturn. Each of the 11 discs (one for each playable character) contains a slideshow of high-resolution CG stills of the character engaged in non-fighting activities such as playing pool or eating ice cream, backed by a Japanese pop song, as well as a karaoke mode.[*][13]

### 3.1.4 Reception

Sega reported pre-orders of 1.5 million units for *Virtua Fighter 2* in Japan, which is nearly as many of the number of Saturns that had been sold in Japan at that point.[*][14] At the time of its release, *Virtua Fighter 2* was the top-selling game for the Saturn, and remains the highest selling Saturn game in Japan with 1.7 million copies.

*Virtua Fighter 2* received generally positive reviews. *Next Generation* gave the game a perfect 5/5 stars, calling it "the ultimate arcade translation" and "the best fighting game ever."[*][24] The magazine cited its "accurate representation of 10 very distinct and realistic fighting styles", "remarkable AI", and "a general attention to detail that sets a new mark for quality game design." [*][25]

*Sega Saturn Magazine* gave the Saturn version a 98%, citing the smooth frame rate, the realistically varied reactions to blows, the huge variety of moves, and the addition of features such as Team Battle Mode.[*][26] Similarly praising the variety of moves and the accuracy of the port, Game Revolution gave the Saturn version an A and concluded that "*Virtua Fighter 2* for the Saturn looks better and smoother than any other polygonal fighting game for the next generation systems. This just might be the best home console fighting game ever." [*][20]

*GamePro*'s Scary Larry called it "the game to own if you have a Saturn", citing the authentic fighting styles and moves, the new modes, the realistic animations with strong attention to detail, and the easy to master controls. He gave it a perfect score in all four categories (graphics, sound, control, and FunFactor).[*][27] The four reviewers of *Electronic Gaming Monthly* felt the port was not as arcade perfect as it could have been, but highly praised the wealth of options and modes, with two of their reviewers declaring it by far the best fighting game on the Saturn thus far.[*][18]

*Game Informer*'s Andy, Reiner, and Paul were slightly more mixed, praising *Virtua Fighter 2* for its depth and variety but criticizing inferior background details in the Saturn port. In addition, Paul felt that the original *Virtua Fighter* required more strategy.[*][19] *Maximum* described the port as "remarkably similar to its coin-op parent - a game that's running on hardware that's 20 times more expensive than the

Sega Saturn." They particularly praised the high-resolution graphics, smooth frame rate, "breathtaking" variety of moves, and the numerous Saturn-exclusive modes and options. With their one criticism being the very vulnerable opponent AI,[*][23] they gave it their "Maximum Game of the Month" award.[*][28]

*GameSpot* gave the PC version a 8.1/10. Praising the game's realism, depth, and opponent AI, and the PC version's inclusion of online multiplayer, they deemed it "unquestionably the best fighting game on the PC, and certainly one of the finest fighting games of all time", adding that the PC version "rivals even the excellent Sega Saturn console port."[*][21]

*Virtua Fighter 2* was ranked as the 19th best arcade game of the 1990s by *Complex*.[*][29]

### 3.1.5   References

[1] http://ysnet-inc.jp/about_e2.html

[2] "Sega unleashes exclusive lineup of arcade hits for Sega Saturn". *Business Wire*. October 30, 1995. Retrieved 2011-05-07.

[3] http://sega.jp/ps2/ages16/

[4] "Preview: Virtua Fighter 2". *GamePro* (68) (IDG). March 1995. p. 20.

[5] http://www.1up.com/features/ disappearance-suzuki-part-1?pager.offset=2

[6] http://www.eventhubs.com/news/2014/oct/25/ top-secret-military-technology-was-used-make-virtua-fighter-2-yep-happened-according-developer/

[7] Leadbetter, Rich (November 1995). "Virtua Fighter: The Second Coming". *Sega Saturn Magazine* (1) (Emap International Limited). pp. 36–41.

[8] "Virtua Fighter 2 Development Diary". *Sega Saturn Magazine* (2) (Emap International Limited). December 1995. p. 46.

[9] Ogasawara, Nob (November 1995). "Sega's Top Guns: An Interview with AM2". *GamePro* (IDG) (86): 28–31.

[10] "It's Almost Here: Virtua Fighter 2". *Maximum: The Video Game Magazine* (Emap International Limited) (1): 116–7. October 1995.

[11] "Protos: Virtua Fighter 2.1". *Electronic Gaming Monthly* (Ziff Davis) (76): 224. November 1995.

[12] Leadbetter, Richard (February 1996). "Virtua Fighter 2 Master Class: Part 1". *Sega Saturn Magazine* (4) (Emap International Limited). pp. 88–91.

[13] "New CG Discs Released". *Maximum: The Video Game Magazine* (Emap International Limited) (2): 131. November 1995.

[14] Hickman, Sam (January 1996). "Virtua Sell Out!". *Sega Saturn Magazine* (Emap International Limited) (3): 7.

[15]

[16]

[17]

[18] "Review Crew: Virtua Fighter 2". *Electronic Gaming Monthly* (Ziff Davis) (79): 31. February 1996.

[19] Reiner, Andrew et al. (January 1996). "Blowout!!!". *Game Informer*. Retrieved 2014-07-15.

[20] "Virtua Fighter 2 Review". *Game Revolution*. Retrieved 13 July 2014.

[21] Kasavin, Greg (October 16, 1997). "Virtua Fighter 2 Review". *GameSpot*. Retrieved 13 July 2014.

[22]

[23] "Maximum Reviews: Virtua Fighter 2". *Maximum: The Video Game Magazine* (Emap International Limited) (3): 140–1. January 1996.

[24] "Platinum Pick: Virtua Fighter 2". *Next Generation* (Imagine Media) **2** (13): 179. January 1996.

[25] "Excellent!". *Next Generation* (Imagine Media) **2** (14): 160. February 1996.

[26] Leadbetter, Richard (December 1995). "Review: Virtua Fighter 2". *Sega Saturn Magazine* (Emap International Limited) (2): 72–73.

[27] "ProReview: Virtua Fighter 2". *GamePro* (IDG) (88): 84. January 1996.

[28] "The Essential Buyers Guide Reviews". *Maximum: The Video Game Magazine* (Emap International Limited) (3): 139. January 1996.

[29] Rich Knight, Hanuman Welch, The 30 Best Arcade Video Games of the 1990s, Complex.com, August 28, 2013.

### 3.1.6   External links

- *Virtua Fighter 2* at the Killer List of Videogames

- *Virtua Fighter 2.1* at the Killer List of Videogames

- *Virtua Fighter 2* at MobyGames

## 3.2 Sega Rally Championship

For the 2007 game also known as *Sega Rally*, see Sega Rally Revo.

***Sega Rally Championship*** is a 1994 arcade racing game developed by AM5[*][2] on the Sega Model 2 board. It was ported over to the Sega Saturn (by AM3) in 1995 and PC in 1997. The unique selling point of *Sega Rally* was the ability to drive on different surfaces (including asphalt, gravel and mud), with different friction properties, with the car's handling changing accordingly. As the first racing game to incorporate this feature, *Sega Rally* is considered to be one of the milestones in the evolution of the racing game genre.[*][3] It was also an early rally racing game and featured cooperative gameplay alongside the usual competitive multiplayer.

The music for the arcade game was composed by Takenobu Mitsuyoshi, while the Sega Saturn port's soundtrack was done by Naofumi Hataya and performed by Joe Satriani.

### 3.2.1 Gameplay

The player can enter a "World Championship" mode consisting of three stages: Desert (which resembles African savanna), Forest (which resembles South American forests) and Mountain (which partly resembles Monaco and Corsica), where their finishing position at the end of one course is carried through to the starting position of the next course. In this mode, it is impossible to reach first-place position by the end of the first track; thus, the player must try to overtake as many opponent cars as possible on each track (while staying within the time limit), and gain the lead over several tracks. If, at the end of the third round, the player is in first place, they are able to play a fourth secret circuit called "Lakeside" (on the Saturn version, this course may then be played in time attack and split-screen multiplayer modes).

Three cars are featured in the game; Didier Auriol's third generation Toyota Celica GT-Four and Juha Kankkunen's Lancia Delta HF Integrale which are both available from the start, and Sandro Munari's Lancia Stratos HF which is unlocked by finishing Lakeside in first place in home versions of the game.[*][4] Players are given the option to drive each car in either manual or automatic transmission.[*][5]

### 3.2.2 Development

*Sega Rally Championship* was directed by Kenji Sasaki, a former Namco employee known for his work on *Ridge Racer*. Seeking to develop a racing game that was distinct from the popular arcade titles *Ridge Racer* and *Daytona USA*, Sasaki chose the rally racing subgenre, which he felt was "taboo" in the Japanese gaming community: "We were after something in vogue in terms of motorsport racing and as we were keen on great engine sounds, cool cars and great sensations—the obvious choice was rally." While the game featured only three cars—the Toyota Celica GT-Four, Lancia Delta Group A, and a hidden Lancia Stratos—it was distinguished by its "stylized handling" and some tuning options.[*][1] Asked why the developers chose to use the Celica and Delta, team manager Hiroto Kikuchi answered, "We felt that in the rally, we had to use real rally cars and the chosen vehicles were well known and looked good." Senior programmer Riyuchi Hattori added, "Originally there was talk of using another car from Toyota, but we couldn't find a good one. For example, the Supra would have been just the same as the Celica and not much fun to use in the game, so we ended up with just the one. We also took note of the consumers' opinions, which confirmed that if another car was to be added it should be the Stratos." [*][6] According to game designer Tetsuya Mizuguchi, "we had no experience in driving those cars", but after repeated requests Toyota and Fiat provided feedback for game testing.[*][1] Fiat also made a gentlemen's agreement with the developers allowing the use of official logos and such in *Sega Rally Championship*; there was no formal sponsorship deal for the game.[*][7] Mizuguchi's car was used to produce the in-game sound of the Lancia Delta's engine. While developing the game's visual style, the development team spent three weeks driving from the West Coast of the United States to Mexico, taking photographs for use in texture mapping. At one point, Sasaki became deeply worried about *Rally*'s prospects for success, and even began to question why driving cars was considered "fun". To clear his mind, "I drove up into the mountains with my own car. It was such an enjoyable and exhilarating experience ... This was how the third mountain track in the game was conceived" .[*][1]

The Saturn version of the game had to be almost completely remade, only referencing the graphics of the arcade version, which required detailed planning.[*][1][*][8] Mizguchi recounted, "Our designers went back to the arcade version and worked out the locations, drew pictures and captured the atmosphere and the feeling of distance. Then there was about two weeks discussion on their work. During this time they worked on the car settings and we had Mr. Yoshio Fujimoto, winner in the Toyota Castrol car to advise it. Then Mr. Nakamura, Mr. Hattori, and Mr. Fujimoto went to the Asian Pacific Indonesian Rally for three days and studied the cars." [*][9] Unlike other well-received arcade ports for the Saturn such as *Virtua Fighter 2* and *Virtua Cop*, *Sega Rally Championship* was developed without using the Sega Graphics Library operating system, as it had not yet been completed when work on the game began.[8][10] For similar reasons, a split screen was used for multiplayer mode instead of the Saturn link cable; the developers also felt it was important that multiplayer be available to all owners of the Saturn game, not only those who had also purchased a link cable.[6][10] Finally, the arcade version of *Rally* was designed to be controlled with a steering wheel, and the developers struggled to simulate its drifting techniques using the Saturn's controller.[1] (The game also supports the Saturn Steering Wheel, though it lacks the haptic feedback of the arcade version's steering wheel.[7])

The Saturn version was rushed to the North American market in order to take advantage of the Christmas shopping season.[10] By the time of its release in Japan and Europe, the development team had completed several additional graphical improvements, bug fixes, and front-end options.[10]

### 3.2.3  Release

In Japan, the Saturn port of the game shared the full title of its arcade counterpart, *Sega Rally Championship 1995*; because it was released on December 29, 1995, the year was dropped from the title of the North American and European Saturn ports.[1]

### 3.2.4  Reception

*Sega Rally Championship* was met with almost universally positive reviews. On release, *Next Generation* scored the Saturn version of the game 5/5 stars, praising its "down-and-dirty feel", "truly phenomenal high-speed visuals", and "quick, responsive control."[11] The magazine cited the game's physics and handling as "nothing short of remarkable".[12] *Game Informer*'s Reiner and Andy gave *Sega Rally* scores of 8/10 and 8.5/10, making note of technical improvements over the Saturn version of Sega's *Daytona USA*, which Andy nonetheless felt was the better game. *Game Informer*'s Paul was more effusive, rating the Saturn port 9.25/10 for its "far better racing feel" and superior graphics to *Daytona*.[13] *Sega Saturn Magazine* gave the game a 97%, praising the difficulty of unlocking the secret course and secret car, and remarking that "whilst there's enough drag, slide action and difficulty wrestling with the controls to convince you the programmers know what it's like to drive a rally car, there's never so much realism that you'd have to know how to drive one yourself to play the game."[14] Both of the sports reviewers for *Electronic Gaming Monthly* gave the Saturn version an 8.5 out of 10, saying it "has all of the action and adventure of its arcade cousin. If you were disappointed with *Daytona*, you won't be with Sega Rally."[15] Bruised Lee of *GamePro*

praised the additional features of the Saturn version and technical improvements over *Daytona USA*, but criticized that the sounds, while identical to the arcade version, are unexciting compared to other racing games. He said that the power-slide technique can be initially frustrating but once mastered is "effective and fun."[16] *Maximum* decreed the conversion to be "every bit as good as anyone could have ever hoped", stating that aside from the frame rate being reduced to 30 frames per second, it is essentially identical to the arcade version. They also complimented the inclusion of a two-player mode, numerous options, and secret modes, and gave it five out of five stars.[17]

*Sega Rally* was named the best racing game of all time by *Retro Gamer* magazine, which ranked it at the top of its "Top 25 Racing Games Ever" list.[18] Codemasters have cited *Sega Rally* as a strong influence on their first *Colin McRae Rally* game.[19] In *Guinness World Records: Gamer's Edition 2009* the Saturn version of the game made it to 44th position in the list of the Top 50 Console Games, due to its "distinct handling style and superb track design."[3] IGN staff writer Levi Buchanan ranked *Sega Rally Championship* 6th in his list of the top 10 Sega Saturn games, saying "Yes, the Dreamcast version is much better and the current-gen sequel... is stunning, but this Saturn arcade port was one of the top reasons to stick by SEGA as it flailed through the 32-bit days."[20]

### 3.2.5  References

[1]  "The Making Of: *Sega Rally Championship 1995*". *Edge*. 2009-10-02. Archived from the original on November 29, 2014. Retrieved 2014-11-14.

[2]  Sega Arcade Developers

[3]  *Guinness World Records: Gamer's Edition 2009*, page 103.

[4]  "Exclusive!! Blue Stratos". *Sega Saturn Magazine* (Emap International Limited) (3): 42. January 1996.

[5]  "Hot at the Arcades". *GamePro* (IDG) (69): 18. April 1995.

[6]  "AM3: The Director's Cut!". *Sega Saturn Magazine* (Emap International Limited) (4): 44–45. February 1996.

[7]  "Sega Rally Short Stories". *Maximum: The Video Game Magazine* (Emap International Limited) (1): 137. October 1995.

[8]  "AM3 Speak!". *Sega Saturn Magazine* (Emap International Limited) (3): 43. January 1996.

[9]  "AM3 Interview: The CS Conversion Team Exposed to Maximum Shock!". *Maximum: The Video Game Magazine* (Emap International Limited) (2): 40. November 1995.

[10] "Sega Rally: AM3's Awesome Arcade Racer Is Superlative on Sega Saturn!!". *Maximum: The Video Game Magazine* (Emap International Limited) (3): 6–13. January 1996.

[11] "Sega Rally Championship". *Next Generation* (Imagine Media) **2** (13): 178. January 1996.

[12] "Top Gear". *Next Generation* (Imagine Media) **2** (14): 160. February 1996.

[13] Reiner, Andrew et al. (January 1996). "Easy Left, Baby". *Game Informer*. Retrieved 2014-07-15.

[14] Automatic, Rad (January 1996). "Review: Sega Rally". *Sega Saturn Magazine* (Emap International Limited) (3): 78–79.

[15] "Box Score: Sega Rally Championship". *Electronic Gaming Monthly* (Ziff Davis) (78): 188. January 1996.

[16] "ProReview: Sega Rally Championship". *GamePro* (IDG) (89): 58. February 1996.

[17] "Maximum Reviews: Sega Rally". *Maximum: The Video Game Magazine* (Emap International Limited) (3): 142. 1996.

[18] "Top 25 Racing Games... Ever! Part 2". *Retro Gamer*. 21 September 2009. pp. 5–6. Retrieved 2014-01-18. The Saturn version of Sega Rally was truly astounding, a real showcase of the brilliance of the machine. The peerless arcade port would encapsulate everything that was wonderful about the arcade game. The tense two-player dashes, the racing refinement by you as a player to unlock the Stratos and to continually return to it so you could shave a few more seconds off your best time – because you always knew it was possible. Sega has always proven to be the flag bearer of videogame exhilaration – something that is so governing in the racing genre – and Sega Rally is perhaps the finest testament to that notion.

[19] Edge Staff (February 5, 2010). "The Making Of: Colin McRae Rally". Edge. Archived from the original on 3 April 2013. The basic premise for the game was based around the car handling in Sega Rally," confirms Guy Wilday, producer of the first four CMR games. "Everyone who played it loved the way the cars behaved on the different surfaces, especially the fact that you could slide the car realistically on the loose gravel. The car handling remains excellent to this day and it's still an arcade machine I enjoy playing, given the chance.

[20] Buchanan, Levi (2008-07-29). "Top 10 SEGA Saturn Games". *IGN*. Retrieved 2013-04-03.

## 3.3 The House of the Dead (video game)

**The House of the Dead** is a first-person, light gun arcade game, first released by Sega in Japan in September 13, 1996, with the international released following in March 4, 1997.

Players assume the role of agents Thomas Rogan and "G" in their efforts to combat the products of the dangerous, inhumane experiments of Dr. Curien, a mad scientist.

### 3.3.1 Gameplay

*The House of the Dead* is a rail shooter light gun game. Players use a light gun (or mouse, in the PC version) to aim and shoot at approaching zombies. The characters' pistols use magazines which hold 6 rounds; players reload by shooting away from the screen. A set of torches next to the magazine of each player represents remaining health. When a player sustains damage or shoots a civilian, one of their torches is removed. The player dies when all torches are lost. First-aid packs are available throughout the game which restore one torch. These are found either in the possession of civilians whom the player has rescued or inside breakable objects. Similarly, there are also special items located in breakable objects that will grant a bonus to whoever shoots it.

Throughout the course of the game, players are faced with numerous situations in which their action (or inaction) will have an effect on the direction of gameplay. This is exemplified in the opening stage of the game when a civilian is about to be thrown from the bridge to his death. If the player saves the civilian, they will enter the house directly through the front door; however, if the player fails to rescue the civilian, the character is redirected to an underground route through the sewers. If the player rescues all civilians, a secret room full of lives and bonuses is revealed toward the end of the game.

### 3.3.2 Story

*The Magician*

On December 18, 1998, seven years after the events from House of the Dead: Overkill, AMS Agent Thomas Rogan (who replaced Detective Isaac Washington) receives a distressing, short phone-call from his fiancée, Sophie Richards at the Curien Mansion, the home and laboratory of Dr. Roy Curien, a renowned biochemist and geneticist. Amidst a series of ominous occurrences and disappearances at the mansion, Rogan arrives on the scene with his new partner, Agent G, to immediately discover the estate overrun with hellish creations. A mortally wounded man gives them a small field journal showing information about all of Curien's deadly creations and their weak points. It is used every time the player(s) are confronted by a boss.

It is revealed that Curien was obsessed with discovering the nature of life and death. While supported by the DBR Corporation and its own team of scientists, Curien's relentless pursuit of this goal slowly drove him insane, with his behavior growing more erratic and the nature of his experiments beginning to take a gruesome turn. Curien's plan for his research ultimately resulted in the release of his experimental subjects free into his mansion. Wasting no time, Rogan and G storm inside the mansion in order to find and save Sophie, as well as the several other scientist trapped inside, where they witness first-hand the terror unleashed by Curien's zombies and abominations.

When Rogan and "G" arrive at the mansion, they find Sophie, but she is just as soon captured by a gargoyle-bat like abomination called the Hangedman, who takes her away to the mansion. They later find Sophie in an empty room, but she is knocked against a wall by Chariot, a gray armored supersoldier carrying a blood stained halberd. Rogan and "G" defeat Chariot and attend to Sophie, who seemingly succumbs to her injuries. Rogan, in a fit of rage, goes off to avenge his fiancée by seeking out Hangedman. He and G find him on the rooftops surrounding the courtyard, where he intends to stop them. He then drops two scientists to their deaths before fighting Rogan and G. He nearly kills them by knocking them off the roof (they hang on to the edge as they continue to fight.) They finally shoot him down, causing him to fall to his death.

The two push on to find Dr. Curien, while having to fight an even larger horde of zombies in the process. The eventually reach him, but he escapes into his underground laboratory and releases the Hermit (a giant spider crab) to finish them. They manage to kill it, and continue the chase.

Upon confronting Curien a second time, the AMS agents are introduced to his masterpiece, The Magician, a humanoid demon-esque creature that possesses a mastery of fire. After Curien releases the creature from its incubation chamber, the Magician reveals itself to be self-aware, refusing to serve any master; Dr. Curien is subsequently killed by his creation. To prevent the Magician from escaping the mansion and destroying the world, Rogan and 'G' confront it in one final battle. Before dying, the Magician gives one last, chilling warning, and then explodes. Rogan and 'G' leave the mansion, taking one last look at it from the outside.

There are, however, alternate endings that the player can achieve upon completing the game. One ending, in which the camera pans to the foyer one last time; the doors open, revealing Sophie to be alive, running towards the camera saying "Thank you!", showing that her injury from earlier did not kill her. Another shows Sophie has now become one of the undead—the last corpse remaining, all depending on the player's final score and the number of continues used.

### 3.3.3 Characters

- **Thomas Rogan**: A young trained AMS agent who arrives at the Curien Mansion to investigate a series of recent disappearances, alarming events and rescuing missions at the Curien Mansion. His fiancée is Sophie Richards, from whom he receives a distressed phone call, is an employee at the mansion. He, along with Agent G, must hurry to save Rogan's fiancée and, in the process, uncover evidence of depraved scientific endeavor and soon become attestors to the birth of a horrifying evil that must be stopped from leaving the mansion. The character's last name was misspelled as "Rowgun" on the arcade cabinet.

- **Agent "G"**: An AMS agent and Rogan's partner, who accompanies Rogan to the Curien Mansion. If players choose him as the first character, the in-game dialogue in the scene will change, for example, Sophie will address him differently upon arrival at the mansion.

- **Dr. Roy Curien**: An acclaimed biochemist and geneticist who worked for the DBR Corporation and the main antagonist of the game. He was responsible for *The Curien Mansion Incident*. He was obsessed with discovering the very nature of life and death, which eventually drove him mad. His increasingly questionable methods and experiments garnered the suspicion and alarm of his colleagues, until it was too late. His deteriorating mental state culminated in the creation and wanton release of hideous monsters from the laboratory to the mansion and surrounding estate. Thanks to the efforts of Rogan and G, the creatures were prevented from escaping the mansion grounds.

- **Sophie Richards**: Rogan's fiancée,(After)Beloved Wife. She tried to call Rogan for help upon onset of the disaster. She managed to escape the mansion, but was somehow knocked unconscious and is found by

Rogan lying in front of a fountain in the front courtyard. She awakes and runs for Rogan/G but is captured and brought back into the house by *The Hanged Man*. She is later found in a large room inside the mansion, but is severely wounded by *The Chariot*. Her survival depends on the rating the player receives upon completion of the game. If the player gets 60,000 score or above, the good ending scene appeared. If below 60,000, she is the undead - the last corpse. While the player's performance would determine her survival in the first game, canonically she survived, married, and had a daughter with Rogan, whom they named Lisa Rogan.

- **Daniel Curien**: The mysterious son of Dr. Roy Curien.

**Sentient Mutants**

- **The Magician (Type 0)**: Dr. Curien's ultimate creation, a humanoid demon with mastery over fire and powers of levity and anti-gravity propulsion. Realizing that it need not take instruction from inferior beings and that it is more powerful than its creator, it disobeys Curien and destroys him with a fireball. It is defeated at the hands of Agents Rogan and 'G'; with its final breath it utters one final, ominous threat to the agents: a grim harbinger of future events. **"You...haven't seen...anything yet!"** before exploding violently in a cataclysmic fireball before the winter sunset. It would be reanimated once more just over a year later by Caleb Goldman as the penultimate boss during the events of The House of the Dead 2 to oversee the culmination of Curien's unfinished research and safeguard Goldman's magnum opus while it goes to completion.

- **The Hanged Man (Type 041)**: A semi-sentient gargoyle-like Beastman who controls the 'Devilions', mutated bats who also did not escape the scourge of Curien's research. It is the secondary antagonist in the game who is encountered before the truth behind Dr. Curien's research becomes apparent as the agents travel deeper into the mansion. *The Hangman*s DNA would also be utilized by Caleb Goldman to create Zeal in *The House of the Dead 2*; *The Hanged Man* is, however, only sentient to the extent that he can speak; he therefore remains servile to his masters, unlike *The Magician*.

### 3.3.4 Reception

*The House of the Dead* garnered generally positive reviews, the arcade version being held in the highest regard with Allgame awarding it 4.5 out of 5 stars.[3] However, the Saturn and PC versions gained slightly less praise due to their lack of polish. GameRankings gave it a score of 63% for the PC version[2] and 70.54% for the Saturn version.[1]

When Indianapolis attempted to ban violent video games it argued that *The House of the Dead* was obscene and so unprotected by the First Amendment. This required U.S. Appeals Court Judge Richard Posner to review the game at length, ultimately finding Indianapolis' ban was unconstitutional. Unimpressed by the graphics, Judge Posner wrote "The most violent game in the record, "The House of the Dead," depicts zombies being killed flamboyantly, with much severing of limbs and effusion of blood; but so stylized and patently fictitious is the cartoon-like depiction that no one would suppose it "obscene" in the sense in which a photograph of a person being decapitated might be described as "obscene." It will not turn anyone's stomach."[13]

### 3.3.5 Ports

The game was ported in 1998 to Sega Saturn by Tantalus, and to Windows (PC-CD) by Sega. The conversion suffered from somewhat rushed development.[14] Official Sega Saturn Magazine criticized the graphics and framerate of the game, which ran at 20 frames per second. However, extra game modes were added to the port which include a home specific mode that allows the player to select a character; and also a boss battle mode, which pits the player against the game bosses back to back.

### 3.3.6 Further reading

- "*The House of the Dead*". *EGM²*. June 1997.

### 3.3.7 External links

- *The House of the Dead* on MobyGames

- *The House of the Dead* at the Killer List of Videogames

- *The House of the Dead* at SegaSaturn.co.uk

### 3.3.8 References

[1]　"The House of the Dead for Saturn". GameRankings. Retrieved 2014-02-16.

[2]　"The House of the Dead for PC". GameRankings. Retrieved 2014-02-16.

[3]　Baize, Anthony. "The House of the Dead (ARC) - Review". Allgame. Retrieved 2014-02-16.

[4]　House, Matthew. "The House of the Dead (PC) - Review". Allgame. Retrieved 2014-02-16.

[5]　House, Michael L. "The House of the Dead (SAT) - Review". Allgame. Retrieved 2014-02-16.

[6]　Randell, Kim (1998). "PC Review: House of the Dead". *Computer and Video Games*. Archived from the original on 2007-06-24. Retrieved 2014-02-16.

[7]　Edge staff (April 1998). "House of the Dead (SAT)". *Edge* (57).

[8]　"The House of the Dead (SAT)". *Game Informer* (61). May 1998.

[9]　Ferris, Duke (September 1998). "The House of the Dead Review (SAT)". Game Revolution. Retrieved 2014-02-16.

[10]　Fielder, Joe (1998-04-23). "The House of the Dead Review (SAT)". GameSpot. Retrieved 2014-02-16.

[11]　Williamson, Colin (December 1998). "House of the Dead". *PC Gamer*. Archived from the original on 2000-03-03. Retrieved 2014-03-26.

[12]　"PC Review: The House of the Dead". *PC Zone*. 1998.

[13]　*American Amusement Machine Ass'n v. Kendrick*, 244 F.3d 572 (7th Cir. 2001).

[14]　https://archive.org/stream/Official_Sega_Saturn_Magazine_031/Official_Sega_Saturn_Magazine_031_-_may_1998_UK#page/n63/mode/2up

## 3.4 Panzer Dragoon Saga

*Panzer Dragoon Saga* (アゼルパンツァードラグーン RPG *Azel: Panzer Dragoon RPG*) is a role-playing video game developed by Sega subsidiary Team Andromeda and released for the Sega Saturn in 1998. It is the third game released in the *Panzer Dragoon* series (not including *Panzer Dragoon Mini*, a spin-off title for the Sega Game Gear), and is the only game in the series that is not a rail shooter. The player controls a young hunter named Edge as he attempts to free the world from the will of the Ancients with the help of a powerful flying dragon, which can fire lasers from its mouth. The game blends strategic semi-realtime combat with free-roaming exploratory sequences while riding and on foot.

Upon release, *Panzer Dragoon Saga* received critical acclaim. It was the last *Panzer Dragoon* game developed by Team Andromeda before the team disbanded, after which several of its members went on to Smilebit and later developed the fourth game in the series, *Panzer Dragoon Orta*, for Xbox. The game has not been re-released on any other platforms, despite desire from fans.*[1] In 2009, director Yukio Futatsugi stated that the game's source code no longer existed in Sega's records, which indicates that more work than normal would be required to port the game to modern platforms.*[2]

### 3.4.1 Gameplay

The player, as Edge, explores 3D environments on foot or riding the dragon, solving puzzles and interacting with non-player characters to further the story. The player can use a cursor to "lock on" to items of interest; locking onto distant characters lets Edge eavesdrop on their conversation. The dragon is used to explore the game's overworld and dungeons. Unlike the other *Panzer Dragoon* games, the dragon's course is not "on rails", and can be moved in any direction.

**Battle system**

*Panzer Dragoon Saga* uses a random encounter system to trigger battle sequences, which take place in mid-air on the dragon. The player has three *action gauges* that deplete with each move made. Basic moves such as firing Edge's gun or unleashing a laser attack from the dragon cost one gauge; special attacks (called "berserks") may use more. The gauges refill in real time.

The player can circle the enemy to target weak points and escape dangerous positions. Edge's gun is used to concentrate fire on individual targets, and can be upgraded with attachments; for example, the Sniper attachment increases damage done to weak points. The dragon's homing lasers attack multiple targets.

The player can "morph" the dragon to boost its attack, defense, agility and berserk attributes; boosting one attribute depletes another. Additional morph forms are acquired by collecting hidden "D-Unit" items. After battles, the player earns experience points that boost the dragon's power and ability.

*Edge and his dragon in combat. Note the combat menu on the left, the safety radar in the center and the three action gauges, one depleted, on the right.*

### 3.4.2 Plot

*Panzer Dragoon Saga* is set after *Panzer Dragoon II Zwei* and before *Panzer Dragoon Orta*.

Edge is an imperial soldier guarding an excavation site in a post-apocalyptic environment where artefacts from a lost civilization are being recovered. After helping fend off an attack from one of the wild mutant creatures that roam the land, Edge discovers the body of a young woman buried in a wall. The site is attacked by a faction of mutinous Imperial troops; its leader, Craymen, seizes the girl and shoots Edge and his companions. Edge falls into an underground reservoir but is mysteriously unharmed. A flying dragon descends into the reservoir and seems to communicate with Edge; with its help, Edge escapes the cave and swears revenge on Craymen.

### 3.4.3 Development

Development for *Panzer Dragoon Saga* began around the same time as its predecessor, Panzer Dragoon II Zwei, and its development group started as an offshoot of that game's team. They eventually grew to around 40 members, twice as many as that of the team working on Panzer Dragoon Zwei. The team wanted to create a game of a different genre to others in the *Panzer Dragoon* series but retained the series' identity.[*][3]

Like its predecessor, the 3D software Softimage was used in development. The game was listed in the top five most anticipated titles on Japan's *Sega Saturn Weekly* magazine for two years before its release.[*][1]

Even as the Sega Saturn faltered in the console market, Team Andromeda struggled with their own difficulties in developing the game. While they appreciated the creative freedom they had in attempting to produce gameplay that was unique, there was a need to adhere somewhat to the standard conventions of the RPG genre, as well as to incorporate the aerial shooting elements of the series. Eventually, they achieved their goal as they developed a well-received battle system that utilized both turn-based and real-time combat, and allowed the player to alter the dragon and its abilities on-the-fly, as opposed to the specific evolution paths of the previous game.

Team leader Yukio Futatsugi has stated that this was done so as to allow for more combinations to compensate for the lack of characters, as most RPGs usually allow the player to control a party of several members, as opposed to the single character and his dragon in *Panzer Dragoon Saga*.[*][1]

The game also intentionally eschewed the highly populated worlds of most RPGs, as Futatsugi felt the lack of NPCs lent a sense of loneliness to the game. This is further evidenced by his statement that the game could only have been done on the Saturn, rather than the Sony PlayStation, due to the former's more somber color palette, which served to further highlight the game's desolate, post-apocalyptic atmosphere.[*][1]

Team Andromeda also had to deal with the tragic losses of a couple of their members during the game's development: one to a motorcycle accident, and another to suicide. The latter occurred during development of the game, and not, as is often rumored, due to the eventual poor sales of the game.[*][1]

### 3.4.4 Release

*Panzer Dragoon Saga* was one of the last games released for the Saturn outside Japan.[*][4]

Coming late in the Sega Saturn's life, the game was released in very small quantities in the US and Europe. Only 6,000 copies were first produced for the game's American launch in May 1998, and many retailers failed to meet pre-order demand. Sega released a second batch of 12,000 copies the following June and then another 12,000 in the late summer.

In May 2009, online game-downloading service GameTap's general manager Sene Sorrow stated they had the rights to publish the game, but didn't believe there was enough demand to make it a priority.[*][5] Because of the title's limited print run, English-language copies of *Panzer Dragoon Saga* are rare. What copies do exist tend to be recognized as valuable by their owners and, as a result, copies tend to sell for a relatively high price on eBay, often raking in over $300 US.[*][6] Although the game is frequently requested to be re-released on a modern format, thus far the game remains a Saturn-only release. Team leader Yukio Futatsugi has also confirmed that the original source code for the game has been lost, adding further weight to the unlikelihood of a port.*[2] There is now a demand from fans for the source code to be recovered.*[1]

### 3.4.5 Legacy

*Panzer Dragoon Saga* was received with unanimous praise from international gaming publications, citing its sophisticated art direction, vivid and unusual story, and its cinematic, fluid battle system as particularly noteworthy. Official *Sega Saturn Magazine* UK gave it a review score of 96%, and provided the entire first disc of the game with a £4.99 edition.

In addition to its persistent fan following, *Panzer Dragoon Saga* continues to be well regarded by critics and was featured in IGN.com's list of the top 100 games of all time in 2007*[7] and in G4's top 100 games of all time in 2012.*[8] On Gamerankings.com, the game is ranked as the most critically acclaimed Saturn game of all time, with an aggregate score of 92.87%.

### 3.4.6 Soundtrack

A "Mini Album" was released prior to the official release of the game itself, a two-disc set with the first CD containing a selection of music from the game and the second disc acting as a game demo. The two-disc OST was released later with a limited print run. It would be re-released in 2001 as "Azel: Panzer Dragoon RPG Memorial Album" with two bonus tracks remixing the ending song. However, this print run would also be limited.

Much of the soundtrack was generated in-game using the Saturn version of Invision's Cybersound, as in *Panzer Dragoon II Zwei*.

### 3.4.7 See also

- List of dragon video games

### 3.4.8 References

[1] "Panzer Dragoon Saga Sega Saturn Retrospective". *1UP.com*. Retrieved January 22, 2009.

[2] Ciolek, Todd. "Among the Missing: Notable Games Lost to Time". 1UP.com. Retrieved 2014-04-02. In a 2009 interview, *Panzer Dragoon Saga* director Yukio Futatsugi confided that the game's source code no longer existed in Sega's records. While this might have delighted some collectors who paid hundreds on eBay for the game, it's by no means a

death sentence. Even without a game's source code, it's entirely possible for developers to dump the data from a retail copy and emulate it. While the lack of basic code might rule out a heavily reworked version of the game, it simply makes porting *Panzer Dragoon Saga* more work than normal.

[3] "The History of Panzer Dragoon". *GameSpot*. Retrieved September 12, 2007.

[4] Kalata, Kurt. "The History of Panzer Dragoon". *Gamasutra*. Retrieved 21 November 2014.

[5] "GameTap sitting on Panzer Dragoon Saga, Joystiq mobilizes masses". *Joystiq.com*. Retrieved May 13, 2009.

[6] "VideoGamePriceCharts.com". Retrieved 2008-05-12.

[7] "IGN Top 100 Games 2007". *IGN.com*. Retrieved November 24, 2008.

[8] Top 100 Games of All Time: No.22, G4.

### 3.4.9 External links

- The Will of the Ancients A *Panzer Dragoon* series fansite with timelines, theories, fanart, and discussions

- Panzer Dragoon Saga Oasis An extensive fansite dedicated to Panzer Dragoon Saga containing information, creative media, and a full visual guide to the game

- The Art of Panzer Dragoon, A database of official concept art from the series

- RPGamer's *Panzer Dragoon Saga* page

- RPG Fan's Azel Panzer Dragoon RPG OST review

- RPG Fan's Memorial Album review

- Panzer Dragoon Saga Video Project, A site dedicated to *Panzer Dragoon Saga* videos

- Panzer Dragoon Saga Spanish Fan Translation, SEGASaturno & Old Castle Street fan patch

## 3.5 Dragon Force

For the power metal band, see DragonForce.

*Dragon Force* (ドラゴンフォース) is a real-time strategy and tactics video game from Sega created for the Sega Saturn. It was created in Japan and translated for North American release by Working Designs in 1996, a translation that was also used by Sega in Europe under license from Working Designs. A sequel, later translated by fans, was released for the Saturn in Japan in 1998. The first game was re-released for the PlayStation 2 as part of the Sega Ages series.

### 3.5.1 Gameplay

The player assumes the role of one of the continent's eight rulers and sets out to, depending on the ruler, unite the continent, bring peace to the land, and put a stop to the great evil that wants to destroy the land. Each of the eight rulers move along predetermined paths between towns and castles, with castles being the primary objective of the game. When two armies meet, or an army approaches an enemy castle, the focus then switches to that battle.

Gameplay is generally divided into two categories; the strategic "world map" view, and the tactics-oriented battle. On the world map, the player organizes and moves his forces in real time, although the game pauses when the player enters a menu. Armies may only move along predetermined paths between towns and castles.

At the outset of the battle, each side chooses a general and corresponding company of troops to command. After the selection of generals, each side chooses a formation which determines the arrangement of troops. The battle is then fought in real-time, again pausing the action when the player goes into a menu to select commands or use the generals' special attacks or spells. Battles end when one general runs out of hit points or retreats; if both generals' armies are depleted, both generals have the option of dueling or retreating. Generals who run out of hit points are, depending on the general, captured, injured, or (rarely) killed in action. If the player's ruler is defeated in this manner, the player loses the game and must restart from the last save. Once the battle is finished, the process repeats until one army's generals have all been defeated.

Every in-game "week" (a fixed amount of time on the world map), the player attends to administrative duties. During this time, players may give awards to generals (increasing the number of troops they can command or items that increase their capabilities), persuade captive enemy generals to join the player's army, search for items or recruit generals in the ruler's territory, fortify castles, and save the game. Plot-advancing cut scenes frequently take place at the end of the week.

### 3.5.2 Plot

*Dragon Force* is set in the world Legendra, which lived in an era of prosperity under the watch of the benevolent goddess Astea, until it came under siege by the evil god Madruk and his armies. To stop him, a defender came in the form of the Star dragon Harsgalt, and his chosen warriors known as the Dragon Force. Unfortunately, personal disputes amongst the Dragon Force led to their downfall and left Harsgalt to face Madruk. The two faced each other in a fight to death, and Harsgalt, unable to kill Madruk, sealed him away until

eight new chosen warriors could rise to permanently defeat him.

300 years later, the seal imprisoning Madruk has weakened and two of his Dark Apostles, Scythe and Gaul, have begun working towards his release. To ensure none would stop their master, the two of them manipulate the eight nations of Legendra into warring amongst themselves. Eventually, one of the monarchs will successfully end the war, though the events of how it occurs vary depending on the monarch. Regardless, each of the monarchs will discover that they are the eight members of the Dragon Force, and that the only way they can kill Madruk is by obtaining the Dragon Power left by Harsgalt.

Despite attempts to stop them by Scythe and Gaul, whichever monarch the player controls gains the power, and then has to use it to defeat Madruk's final apostle, a robot named Katmondo. Subsequently, Madruk's prison continues to weaken, allowing him to release his army of dragonmen. Despite his army's release, the Dragon Force fight their way to Madruk's prison and find his three Dark Apostles waiting for them there. Whichever monarch that has the Dragon power leaves to face Madruk, while the remaining seven fight the Dark Apostles, and defeat them despite the three becoming even more powerful thanks to the seal on Madruk weakening. The monarch with the Dragon Power then faces and kills Madruk, finally ending his threat. Though the monarch's generals initially lose hope of them surviving, they are saved by Astea, who leaves the world to be governed by the mortals, saying it is time for them to stand on their own. Whatever events that follow during the credits vary depending on the monarch the player uses.

Within the game, eight different storylines exist—one for each monarch. The campaigns for Goldark and Reinhart can only be accessed after the game has been completed, as they contain spoilers from the outset.

### 3.5.3 Reception

*Dragon Force* won *Electronic Gaming Monthly's* Game of the Month award, and its Saturn Game of the Year award for 1996. *EGM* later ranked the game at #111 on its list of 'The Greatest 200 Videogames of Their Time'.[*][5]

### 3.5.4 Sequel

*Dragon Force II: Kamisarishi Daichi ni* was developed and published by Sega, and released only in Japan in 1998.

### 3.5.5 References

[1] http://www.gamerankings.com/saturn/
197149-dragon-force/index.html

[2] *Electronic Gaming Monthly*, issue 90 (January 1997), page
60

[3] http://www.gamespot.com/reviews/dragon-force-review/
1900-2533911/

[4] *Electronic Gaming Monthly*, issue 92 (March 1997), pages
82-88

[5] The Greatest 200 Videogames of Their Time from 1UP.com

### 3.5.6 External links

- Official website (PS3 version)

- *Dragon Force* at MobyGames

## 3.6 PowerSlave

For other uses, see Powerslave (disambiguation).

*PowerSlave*, known as *Exhumed* in PAL territories and
*A.D. 1999: Pharaoh's Revival* (西暦１９９９ファラ
オの復活) in Japan, is a first-person shooter developed
by Lobotomy Software and published by Playmates Interactive Entertainment.  It was released in the U.S, Japan,
and Europe for the Sega Saturn, Sony PlayStation, and
DOS over the course of a year from late 1996 to late
1997.*[2]*[3]*[4] On May 24, 2015, *Powerslave EX*, an
unofficial remake based on the PlayStation version, was released for free.*[1]*[5]

### 3.6.1 Synopsis

*PowerSlave* is set in an area around the ancient Egyptian city
of Karnak in the late 20th century. The city has been seized
by unknown forces, with a special crack team of hardened
soldiers sent to the valley of Karnak, to uncover the source
of this trouble. However, on the journey there, the player's
helicopter is shot down and the player barely escapes. The
player is sent in to the valley as the hero to save Karnak
and the World. The player finds himself battling hordes of
evil creatures known as the Kilmaat, including mummies,
Anubis soldiers, scorpions, and evil spirits.  The player's
course of action is directed by the spirit of King Ramses,
whose mummy was exhumed from its tomb by these evil
creatures to drain it of its power, and use it to control the
world.

In the console versions, there are two endings, depending on
the player's course of action during the game.  In the first,
good ending, the protagonist has collected eight pieces of
a radio transmission device, so he can send a rescue signal and be extracted from the Valley.  After reclaiming the
mummy of Ramses, the Pharaoh thanks him for his effort,
and promises the player that he will inherit Ramses' Earthly
kingdom, and that the Gods will bless him with eternal life
and make him ruler of the world.  After escaping the col-
lapsing tomb, the player is indeed rewarded as such, and
becomes a powerful and benevolent Pharaoh of the entire
planet.  In the second, bad ending, the player has failed to
collect all eight pieces of the radio transmitter, and is subsequently buried in the tomb of Ramses, only to be excavated
centuries later by the now ruling forces of the Kilmaat.

In the PC version, there are two slightly different endings,
again depending on the player's course of action, but only
in the final stage.  The final stage takes place aboard the Kilmaat mothership, where a nuclear weapon has been armed
and is set to go off in 15 minutes, and has enough power
to obliterate the planet.  In the bad ending, which occurs if
the player loses all of their lives or fails to disarm the bomb
in time, Earth is destroyed in a massive nuclear blast.  In
the good ending, which occurs if the player makes it to the
bomb on time, the Kilmaat retreat from the planet, but unfortunately for the main character, he's stuck on their ship
and needs to find a way off.

### 3.6.2 Gameplay

Gameplay follows the standard first-person shooter formula.
Familiar elements from the genre, such as collecting keys to
open doors in a level, are present.

The player character must acquire artifacts which give him
new abilities.  Such abilities include being able to jump
higher, levitate, breathe underwater, walk in lava, walk
through force fields and jump further to reach previously
inaccessible areas.  In the console versions, the player goes
from level to level via an overworld map.

### 3.6.3 History

#### Development

At one point, the PC version was to be released by 3D
Realms as one of their games to show off the power of the
Build engine.  During this time, the game was known by
its working title *Ruins: Return of the Gods*. Apogee Software released screenshots of the early working version with
a slideshow of another of its published titles, *Mystic Tow-
ers*. 3D Realms eventually dropped the title, which was then

picked up by Playmates Interactive Entertainment and published.

The U.S title *PowerSlave* is a reference to the Iron Maiden album of the same name, which also features an Egyptian-themed cover.

Voice narration in the game was performed by Don La-Fontaine, also known as "The Movie Trailer Guy".

## Releases

**Console** The first version of game to be released was on the Sega Saturn, shortly followed by a release on the PlayStation, with tweaked gameplay, added architecture, some different levels, and other changes. Both of these versions are based on Lobotomy Software's SlaveDriver engine and feature a true 3D world, similar to *Quake*. The same engine was used to power the Sega Saturn versions of *Duke Nukem 3D* and *Quake*.

Besides some changes in the levels (rooms in one version that are not in another, added architecture in the PlayStation version), the levels Amun Mines, Heket Marsh, Set Palace, Cavern of Peril, and Kilmaat Colony are almost completely different between the two versions. In the Sega Saturn version, ammo and health pick-ups dropped by an airborne enemy remain airborne, as opposed to falling to the ground in the PlayStation version. There are exclusive powerups on the Sega Saturn such as the All-Seeing Eye, Invisibility and Weapon Boost. Also exclusive to the Sega Saturn is the ability to bomb-boost, which is similar to rocket jumping in other FPS games.

Sprites are represented in 2D, similar to games such as *Doom* and *Duke Nukem 3D*. The game features colored dynamic lighting, but only in the console versions.

Level progression is non-linear, letting the player go to any previous unlocked level at any time. Some levels have areas which are only accessible after getting a certain ability or weapon, similar to the *Metroid* series. This adds an exploration aspect not usually seen in FPS games at the time.

Additionally, there are eight pieces of a radio transmission device hidden in eight of the stages. Stages with a hidden transmitter piece will emit a steady beeping noise on the overworld map, and can be heard beeping when the player is near their location. Collecting these pieces will affect the ending of the game.

Exclusive to the Sega Saturn version is a hidden minigame: Death Tank, which is unlocked by collecting all 23 Team Dolls hidden in the game.[*][6]

**PC** The PC version of PowerSlave features many differences from the console versions. The PC version was built on the Build engine, licensed from 3D Realms. The version used is a slightly earlier version of the engine, made sometime before the version used in *Duke Nukem 3D*. The light sourcing from the SlaveDriver engine is not used; the Build engine's own light sourcing is used instead; the game also uses "fake" dynamic lighting where sectors light up as projectiles or "glowing" objects in general pass through.

The HUD interface is different; featuring an ammo counter, lungs (oxygen levels) for swimming and animated mana and blood vessels. Players have usable Mana energy that can cast spells once the spell has been acquired (e.g. collecting a torch allows the player to use energy to illuminate dark areas). Ammo is not universal, instead of blue orbs usable for all weapons, separate ammo is needed. Grenades are used instead of Amun mines. Some sprites are different (e.g. M60 machine gun), sprites are larger and more animated in general. Audible words are used for the player character instead of grunts. Mummies fire a "white skull" attack, or a partly homing red one, that when hit, turns the player into a mummy momentarily, additionally with the most powerful weapon in the game: the Mummy Staff, which can destroy all enemies within range of the player. The player reverts to normal once the weapon is used. Checkpoints are placed throughout the level by indication of golden scarabs and Saving is automatic between levels.

Levels are conducted in a more linear format. Players can replay previously completed levels, but later stages may only be played after completing the level prior. The Manacle of Power fires a lighting cloud above the enemy, rather than firing lighting bolts from the player's hands. Most of the artifacts from the console versions are not present (except the Sobek Mask, which is a spell). The powerups in the Sega Saturn version are included as spells (invisibility, invincibility and double damage). The Ring of Ra weapon is not included. Weapons pause to reload after a certain number of shots fired. Some enemies have different death animations when killed by fire/grenades, bosses have longer death animations. There are extra lives instead of health extensions. The Amnit enemies are not included; instead there is the giant Ammut miniboss which has ramming and biting attacks. There is additional story text. The transmitter, which was a set of eight key items needed in the console versions to get the better ending, is not in the PC version *per se*, but it is seen before the final stage of the game, where the player receives orders to attack the Kilmaat ship. Likewise, the Team Dolls are not in the PC version.

## Re-release

On May 24, 2015, an unofficial remake based on the PlayStation version was released by Samuel "Kaiser" Villarreal for free.[*][1][*][5] In a May 2015 interview, it was announced that publisher Night Dive Studios had acquired the game rights for a digital distribution re-release.[*][5]

### 3.6.4    References

[1] Powerslave EX released, powerslaveex.wordpress.com.

[2] Saturn version release data, GameFAQs.com

[3] PlayStation version release data, GameFAQs.com.

[4] PC version release data, GameFAQs.com.

[5] Ruhland, Perry (May 4, 2015).  "An Interview With the Man Rebuilding Powerslave". techraptor.net.  Retrieved 2015-05-31.

[6] Information on PowerSlave including team doll locations and Death Tank

### 3.6.5    External links

- PowerSlave on MobyGames

- Interview with PowerSlave's Developer

- Review of Saturn Version

## 3.7    World Series Baseball (video game)

This article is about the specific Sega Genesis game.  For other uses, see World Series Baseball (disambiguation).

*Sega Sports' World Series Baseball*, or simply *World Series Baseball*, is a sports game developed by BlueSky Software and published by Sega for the Genesis/Mega Drive and Game Gear.  It is the first game in the series and was originally released in 1994.  A version for the Sega 32X, *World Series Baseball starring Deion Sanders*, would follow in 1995.

The game was a major advancement in Sega baseball games in that it included licensed players and teams, and relatively accurate gameplay.

The series concluded with *World Series Baseball 2K2* on the Xbox.  After that, Sega contracted with 2K Games to take over their sports game contracts and the line continued as the *Major League Baseball 2K* franchise.

### 3.7.1    Reception

*GamePro* gave the Genesis version a rave review, calling it "arguably the best baseball cart ever." They praised the use of real life teams, players, and stadiums, the accurate graphical recreation of the stadiums, the catcher's-eye view of the action, and the generally impressive graphics.[*][2]

### 3.7.2    References

[1] *World Series Baseball* at MobyGames

[2] "Sega Sports' World Series Champ!". *GamePro* (58) (IDG). May 1994. pp. 106–107.

## 3.8    Sega Worldwide Soccer

*Sega Worldwide Soccer* is a series of soccer games by Sega for initially for the Sega Saturn but later was moved to the Dreamcast. They were released between 1995 and the year 2000.[*][1]

### 3.8.1    History

*Sega Worldwide Soccer*, produced by Sega themselves, was a launch game for the Sega Saturn's North American release.[*][2] It was preceded by *International Victory Goal*, one of the debut titles of the console. The game featured international teams and league, play-off and tournament modes. Although it used fictional player names (due to the lack of a license), the non-volatile memory of the Saturn allowed editing of names. The team kits were as close to the official 1996 kits as possible. The game was the top-rated football game until ISS 64 was released one year later. *Worldwide Soccer* was later ported to the PC.

One year later *Sega Worldwide Soccer 98* was released, again for the Saturn. This version featured (still unlicensed) clubs from England, Spain and France, two new stadiums and the same free-flowing gameplay.

One final title, *Sega Worldwide Soccer 2000*, appeared on the Dreamcast. However, instead of being developed in-house, Silicon Dreams (who previously worked with Eidos on the UEFA Champions League series and also World League Soccer) was given the rights to produce a game bearing the Worldwide Soccer name. A *Euro Edition* (capitalizing on the popularity of Euro 2000) was released in Europe.

### 3.8.2 Installments

### 3.8.3 Reception

*GamePro* gave *Worldwide Soccer* a positive review, commenting that "Not since FIFA amazed 3DO owners has another soccer game looked so good and played so well." They particularly praised the graphics, zooming camera, and the demanding gameplay, though they did criticize the "magnetic" dribbling and passing as being less realistic than the dribbling and passing in 3DO FIFA.[*][3]

### 3.8.4 See also

- Wave Master

- List of J. League licensed video games

### 3.8.5 References

[1] "SEGA Worldwide Soccer / Victory Goal series". Mobygames. Retrieved 22 April 2015.

[2] "Sega Hopes to Run Rings Around the Competition with Early Release of the Saturn". *Electronic Gaming Monthly* (Ziff Davis) (72): 30. July 1995.

[3] "Worldwide Soccer Rivals FIFA". *GamePro* (IDG) (83): 78. August 1995.

### 3.8.6 External links

- *Sega Worldwide Soccer '98: Club Edition* at SegaSaturn.co.uk

- *SEGA Worldwide Soccer / Victory Goal* series at MobyGames

## 3.9 Sonic X-treme

Not to be confused with Sonic Extreme.

*Sonic X-treme* was a cancelled platform video game in the *Sonic the Hedgehog* series. Developed by Sega Technical Institute (STI), *X-treme* was designed to capitalize on the success of Sega's mascot character by being the first fully 3D *Sonic* game and the first original *Sonic* title developed for the Sega Saturn. During the course of development, several different styles of gameplay were tried and the plot of the game changed several times.

Originally pitched as a two-dimensional platform game for the Sega Genesis, the game was eventually moved to development on the Saturn and for Windows, intended for release during the holiday season of 1996. However, *X-treme* became stuck in development hell after several incidents, including an unfavorable visit by Sega of Japan executives and issues with acquiring a game engine, made the deadline difficult to achieve. After two of the lead programmers for the project became ill, the game was eventually cancelled. Reviewers and video game journalists have retrospectively considered the possibility of what *Sonic X-treme* could have done for the Saturn had it been released, including comparisons to competing mascot video games *Super Mario 64* and *Crash Bandicoot*.

### 3.9.1 Premise

*Jade Gully Zone, from Senn and Alon's engine*

With the game constantly changing platforms, engines, and development teams, there were many loose storylines in consideration. According to developer Christian Senn, about six or seven story lines were considered during the three-year development timeframe.[*][1] While originally based on the Saturday morning cartoon series,[*][2] the main storyline used in promotion of the final game in magazines involved a Professor Gazebo Boobowski and his daughter, Tiara. The two were the guardians of the six magical Rings of Order, as well as the ancient art of ring-smithing. Gazebo and Tiara feared that Dr. Robotnik was after the six Rings of Order, and called on Sonic to get the Rings before Robotnik could. Dr. Robotnik kidnapped Gazebo after he requested Sonic's help, making it so Sonic had to retrieve both him and the Rings of Order.[*][1] At one point in the development process, there was a possibility for 4 playable characters: Knuckles the Echidna, Tiara Boobowski, Miles "Tails" Prower, and Sonic the Hedgehog.[*][2] Other characters intended to be included in the game were Nack the Weasel and Metal Sonic, who would have been a boss character in the final level. Various moves were added to the characters, such as a ring toss move for Sonic, which was left out of development.[1]

To further the traditional "Sonic formula", every level was designed in a tube-like fashion; Sonic would be able to walk onto walls, thus changing the direction of gravity and the rotation of the level itself, much like the special stages in *Knuckles' Chaotix*. In addition, an unusual, fish-eye lens-styled camera was put into place so players could see more of their surroundings at any given time.[2]

> "3D Sonic is free to move around in a completely open 3D environment. Previously, on the 2D games, things were restricted to a very linear path, whereas now he can run around in the open without any restrictions to his path. The 360-degree rotation allows for new aspects to the gameplay. It means that Sonic can now do things like run from a wall onto the ceiling and explore lots of new hidden areas." [1]
> —Executive Producer Mike Wallis

Senn has highlighted that the 3D gameplay still kept true to the *Sonic* series formula of collecting rings and speeding through game levels. Wallis also made mention to the game's overall layout, consisting of three acts per zone and varying in focuses on speed, exploration, and puzzle solving.[1]

### 3.9.2 Development

Following the completion of *Sonic & Knuckles* in 1994, Sega began working on the next game in its *Sonic the Hedgehog* franchise, which was known in development as *Sonic X-treme* and would have been the first *Sonic* game to feature fully 3D graphics. Development of the game was started by Sega Technical Institute, a U.S.-based developer that had worked on several previous *Sonic* games, beginning on the Sega Genesis and subsequently moving to the Sega 32X.[3] In its earliest conception, *Sonic X-treme* was designed for release on the Sega Genesis as a side-scrolling platform game, much like previous Sonic games for the system.[4] As new consoles and the beginning of the 32-bit era were on the way, the game was later moved to the Sega 32X and was known at this stage under the development names *Sonic 32X*[1] and *Sonic Mars*,[4] after the development name "Project Mars" used for the Sega 32X.[5] Even at this stage, the game's design changed wildly, including concepts such as an isometric viewpoint side-scroller. Eventually, however, development of the game was settled on a full 3D platform game.[4]

As the game's design had changed significantly[4] and evolved beyond the capabilities of the struggling Sega 32X,

the game was shifted again to the Sega Saturn.[6] The Saturn version of the project was initially developed separately by two teams in parallel starting in the second half of 1995. One team, led by designer Chris Senn and programmer Ofer Alon, was in charge of developing the main game for PC.[6][7] The other team, lead by Robert Morgan and including programmer Chris Coffin, worked on porting Senn and Alon's work to the Saturn while developing the "free-roaming, 'arena-style'" 3D boss engine.[6][6] Senn and Alon's "fixed-camera side-scroller" with the ability "to move freely in all directions" was similar to *Bug!*, and featured a fish-eye camera system (called the "Reflex Lens") that gave players a wide-angle view of the action.[4] As a result, levels appeared to move around Sonic.[6]

*Screenshot from Coffin's "boss engine"*

In March 1996, Sega of Japan representatives, including CEO Hayao Nakayama, visited STI headquarters to evaluate the game's progress. They were unimpressed by the main game engine's performance on the Saturn, although Senn and Alon did not have an opportunity to demonstrate the PC version.[6][7] Therefore, Nakayama requested the entire game be reworked around the boss engine.[7] To achieve this in time for the strict December 1996[6] deadline, Coffin's team was moved into a place of isolation from further company politics[6] and worked between sixteen[6] and twenty hours a day.[4] Then, in April, Bernie Stolar approached the STI team and inquired of Wallis what he could do to help the game meet its deadline. Wallis suggested that the game engine from Sonic Team's *Nights into Dreams...* would be helpful. Stolar agreed and acquired the engine. However, the engine's creator and lead programmer of the original Mega Drive Sonic games, Yuji Naka, reportedly threatened to leave the company if it was used.[4][7] STI lost two weeks of development time from the loss of the *Nights* engine.[1]

Senn and Alon had initially continued on with their game engine, undeterred by their work's original rejection, hoping to pitch it to Sega's PC division. However, it was eventually rejected again, prompting Alon to leave Sega.[4][6][7] No part of the STI team worked in unity during development.

> "We had artists doing art for levels that hadn't even been concepted out. We had programmers waiting and waiting and waiting until every minute detail had been concepted out, and we had designers doing whatever the hell they wanted. It was a mess, and because of the internal politics, it was even more difficult to get any work done." [1]
> —Executive Producer Mike Wallis

By August, Chris Senn had become so ill that he was told he had six months to live—though Senn would survive this ordeal—and Chris Coffin came down with a severe case of pneumonia. With both Senn's team and Coffin's team crippled, Wallis was left with an incomplete *X-treme* and only two months before its deadline. At this point, he made the decision to cancel the game.[4] Although Sega initially stated that *X-treme* had merely been delayed,[8] the project was cancelled in early 1997.[7]

### 3.9.3 After cancellation

*With* X-treme'*s cancellation, the Sega Saturn had no original Sonic the Hedgehog platform game released for the console*

With the cancellation of *X-treme*, Sega instead decided to concentrate on a port of the Genesis title *Sonic 3D Blast*, and Sonic Team's *Nights into Dreams...* for the 1996 holiday season.[6] Sonic Team started work on an original 3D *Sonic* title for the Saturn, which eventually became *Sonic Adventure* for the Dreamcast. According to Naka, remnants of the project can be seen in the compilation game *Sonic Jam*.[9][10] STI was officially disbanded in 1996 as a result of changes in management at Sega of America.[3]

For many years, very little content from the game was ever released beyond screenshots that had been released to the

media in promotion of the game prior to its cancellation. However, in 2006 a copy of a very early test engine was sold at auction to an anonymous collector who bought it for US$2500.[11] An animated GIF image of the gameplay was initially released, and the disk image itself was leaked on July 17, 2007 after a fundraising project by the "Assemblergames" website community purchased the disc from the collector.[12] In 2006, Chris Senn opened the "*Sonic X-treme* Compendium" web site and began revealing large amounts of the game's development history to the public, including videos of early footage, a playable character named Tiara, and a large amount of previously unreleased concept music related to the title. He also was given permission by Hirokazu Yasuhara, the level designer for the majority of the original 16-bit *Sonic* titles, to post level designs that were going to be put in the game. Senn, along with the community, announced intentions to recreate the game,[2] but ultimately the project was canceled in January 2010.[13]

In early 2015, fans from *Sonic* fansite "Sonic Retro" obtained the game's source code, polished it up into a playable build, and released it for download on the internet.[14] It is largely faithful to the game's original state, with their changes merely being made to piece together code to make it playable and running on modern computers.[14] The first release only features one level, the Jungle-themed level from the 1996 E3 promotional video, but the team expects to be able to release more levels in the future.[14]

### 3.9.4 Legacy

The *Sonic X-treme* debacle has been cited as a reason for the ultimate failure of the Sega Saturn.[5][15] With the *Sonic the Hedgehog* series being attributed to much of the success of the company's prior system, the Genesis, and Sony and Nintendo both having flagship 3D platformers available early in the life cycle of their consoles (*Crash Bandicoot* and *Super Mario 64*, respectively), Sega was expected by fans to follow suit and produce an official 3D *Sonic* game.[1] With the game's cancellation, the Saturn never did receive an exclusive *Sonic* platform game, but rather only the Genesis port of *Sonic 3D Blast*; *Sonic Jam*, a compilation of the 2D Genesis *Sonic* titles; and *Sonic R*, a racing game. Sonic's debut in a full 3D platform game was not until 1998, with *Sonic Adventure* as a Dreamcast launch title, well after the discontinuation of the Saturn.[2]

Following the game's cancellation, journalists and fans have speculated about the impact a completed *X-treme* might have had on the market. David Houghton of GamesRadar described the prospect of "a good 3D *Sonic* game" on the Saturn as "a 'What if...' situation on a par with the dinosaurs not becoming extinct." [6] IGN's Travis Fahs called *X-*

*treme* "the turning point not only for SEGA's mascot and their 32-bit console, but for the entire company", although he also noted that the game served as "an empty vessel for SEGA's ambitions and the hopes of their fans".*[4] Dave Zdyrko, who operated a prominent website for Saturn fans during the system's lifespan, offered a more nuanced perspective: "I don't know if [*X-treme*] could've saved the Saturn, but ... *Sonic* helped make the Genesis and it made absolutely no sense why there wasn't a great new *Sonic* title ready at or near the launch of the [Saturn]".*[16] In a 2013 retrospective, producer Mike Wallis maintained that *X-treme* "definitely would have been competitive" with Nintendo's *Super Mario 64*.*[7] Websites such as Destructoid and GamesRadar have speculated the game could have been a source of inspiration for future games such as 2007's *Super Mario Galaxy*.*[2]*[6] Several journalists would also note similarities between *X-treme* and the 2013 game *Sonic Lost World*.*[17]*[18]

### 3.9.5    References

[1] Allen, Jonathan. "Spotlight: Sonic X-treme". Lost Levels. Retrieved 2012-07-23.

[2] Davis, Ashley (2008-11-19). "What could have been: Sonic X-treme". Destructoid. Retrieved 2012-07-23.

[3] Horowitz, Ken (June 11, 2007). "Developer's Den: Sega Technical Institute". Sega-16. Retrieved 2014-04-16. **Roger Hector:** When it became obvious that Sony was taking the lead, Sega's corporate personality changed. It became very political, with lots of finger-pointing around the company. Sega tried to get a handle on the situation, but they made a lot of mistakes, and ultimately STI was swallowed up in the corporate turmoil.

[4] Fahs, Travis (2008-05-29). "Sonic X-Treme Revisited - Saturn Feature at IGN". IGN. Retrieved 2012-07-23.

[5] Newton, James (2011-06-23). "Feature: The Sonic Games That Never Were". Nintendo Life. Retrieved 2012-07-23.

[6] Houghton, David (April 24, 2008). "The greatest Sonic game we never got ...". GamesRadar. Retrieved 2012-07-23.

[7] "The Making Of: Sonic X-treme". Edge Online. 2013-04-14. Retrieved 2014-06-15.

[8] "New Sega Happenings". Next Generation Online. Archived from the original on 1996-12-20. Retrieved 2014-05-04. *GunBlade NY* and *Sonic X-treme* have now both been officially scheduled for Saturn release in 1997 ... [*X-treme*] had previously been scrapped to be reworked.

[9] Barnholt, Ray. "Yuji Naka Interview: Ivy the Kiwi and a Little Sega Time Traveling". 1UP.com. Archived from the original on 2014-09-07. Retrieved 2014-03-04.

[10] Towell, Justin (2012-06-23). "Super-rare 1990 Sonic The Hedgehog prototype is missing". GamesRadar. Retrieved 2014-03-04. **Yuji Naka:** The reason why there wasn't a Sonic game on Saturn was really because we were concentrating on NiGHTS. We were also working on *Sonic Adventure*—that was originally intended to be out on Saturn, but because Sega as a company was bringing out a new piece of hardware—the Dreamcast—we resorted to switching it over to the Dreamcast, which was the newest hardware at the time. So that's why there wasn't a Sonic game on Saturn. With regards to *X-treme*, I'm not really sure on the exact details of why it was cut short, but from looking at how it was going, it wasn't looking very good from my perspective. So I felt relief when I heard it was cancelled.

[11] Snow, Blake (2006-03-09). "Man pays $2500 for Sonic X-treme demo". Joystiq. Retrieved 2012-07-23.

[12] McWhertor, Michael (2007-06-04). "Sonic X-Treme "Nights Version"". Kotaku.com. Retrieved 2012-07-23.

[13] Senn, Christian (2010-01-12). "Game Forums @ Senntient.com / Public Announcement Regarding Project-S". Senntient.com. Retrieved 2012-07-23.

[14] http://www.eurogamer.net/articles/2015-02-24-sonic-fans-release-long-lost-tech-demo-of-unfinished-sa

[15] Buchanan, Levi (2009-02-02). "What Hath Sonic Wrought? Vol. 10 - Saturn Feature at IGN". IGN. Retrieved 2012-07-23.

[16] Sewart, Greg (2005-08-05). "Sega Saturn: The Pleasure And The Pain". 1UP.com. Archived from the original on 2012-10-21. Retrieved 2014-03-17.

[17] Ponce, Tony (2013-05-28). "Sonic Lost World trailer reminds me of Sonic X-treme". Destructoid. Retrieved 2013-08-01.

[18] Sliwinski, Alexander (2013-05-28). "Sonic: Lost World finds gameplay footage". Joystiq. Retrieved 2013-08-01.

## 3.10    List of Sega Saturn games

*Second model Japanese Sega Saturn*

The **Sega Saturn** (セガサターン *Sega Satān*) is a 32-bit fifth-generation video game console developed by Sega and released on November 22, 1994 in Japan, May 11, 1995 in North America and July 8, 1995 in Europe as the successor to the successful Sega Genesis. At the center of the Saturn is a dual-CPU architecture and a total of eight processors. Its games are in CD-ROM format, and its game library contains a number of arcade ports as well as original titles.

Development of the Saturn began in 1992, the same year Sega's groundbreaking 3D Model 1 arcade hardware debuted. The system adopted parallel processors before the end of 1993, and was designed around a new CPU specially commissioned by Sega from Japanese electronics company Hitachi. When Sega learned the full capabilities of the forthcoming Sony PlayStation console in early 1994, the company responded by incorporating an additional video display processor into the Saturn's design. Successful on launch in Japan due to the popularity of a port of the arcade game *Virtua Fighter*, the system debuted in the United States in a surprise launch four months before its scheduled release date, but failed to sell in large numbers. After the launch, Sega's upper management structure changed with the departures of chairman David Rosen and Sega of Japan CEO Hayao Nakayama from their roles in the American division, and Sega of America CEO Tom Kalinske from the company altogether. This led to the additions of Shoichiro Irimajiri and Bernie Stolar to Sega of America, who guided the Saturn to its discontinuation in 1998 in North America, three years after its release. The Saturn's complex system architecture resulted in the console receiving limited third-party support, which inhibited commercial success. The failure of Sega's development teams to create a game in the *Sonic the Hedgehog* series, known in development as *Sonic X-treme*, has also been attributed as a factor in the console's poor performance.

After the launch of the Nintendo 64 by Nintendo in late 1996, the Saturn began losing market share rapidly in the United States, and company management began to publicly distance itself from the system. By March 1998 the Saturn had sold 9.5 million units worldwide, significantly fewer than the sales of its biggest rival, the PlayStation. The Saturn's installed base reached over 5 million units in Japan, over 2 million units in the United States, and over 970,000 units in Western Europe. It is considered a commercial failure, contributing heavily to the loss of US$309 million for Sega by 1998, and another $450 million during 1998. Reception to the Saturn is mixed based on the console's game library and complex internal hardware. Sega's management has also been criticized for its decision-making during the system's development and cancellation.

This is an incomplete **list of video games (597)** released for the Sega Saturn video game console. This list is organized initially in alphabetical order, but it can also be organized by developer, publisher, or year of release.

### 3.10.1 Games

Contents

- 0–9
- A
- B
- C
- D
- E
- F
- G
- H
- I
- J
- K
- L
- M
- N
- O
- P
- Q
- R
- S
- T
- U
- V
- W
- X
- Y
- Z
- References

### 3.10.2 See also

- List of Sega video game franchises

- Lists of video games

### 3.10.3 References

[1] "Sega Saturn - Games". Allgame. Retrieved March 4, 2014.

[2] "Sega Saturn games" (in Japanese). Sega of Japan. Retrieved March 4, 2014.

[3] "Sega Saturn games - Licensees (1994-1995)" (in Japanese). Sega of Japan. Retrieved March 4, 2014.

[4] "Sega Saturn games - Licensees (1996)" (in Japanese). Sega of Japan. Retrieved March 4, 2014.

[5] "Sega Saturn games - Licensees (1997)" (in Japanese). Sega of Japan. Retrieved March 4, 2014.

[6] "Sega Saturn games - Licensees (1998-2000)" (in Japanese). Sega of Japan. Retrieved March 4, 2014.

[7] "Satakore - Sega Retro". 13 August 2014. Retrieved 23 November 2014.

# Chapter 4

# Further Reading

## 4.1 History of video game consoles (fifth generation)

The **fifth-generation era** (also known as the **32-bit era**, the **64-bit era** and the **3D era**) refers to computer and video games, video game consoles and video game handhelds from approximately 1993 to 2003.[*][1] For home consoles, the best-selling console was the PlayStation by a wide margin, followed by the Nintendo 64 and then the Sega Saturn. For handhelds, this era was characterized by significant fragmentation, because the first handheld of the generation, the Sega Nomad, had a lifespan of just two years, and the Virtual Boy had a lifespan of less than one. Both were discontinued before the other handhelds made their debut. Nintendo's Game Boy Color was the winner in handhelds by a large margin. There were also two updated versions of the original Game Boy: Game Boy Light (Japan only) and Game Boy Pocket.

The era is known for its pivotal role in the video game industry's leap from 2D computer graphics to 3D computer graphics, as well as the shift from home consoles using ROM cartridges to optical discs. The development of the Internet also made it possible to store and download tape and ROM images of older games, eventually leading 7th generation consoles (such as the Xbox 360, Wii, PlayStation 3, PlayStation Portable, and Nintendo DSi) to make many older games available for purchase or download. This generation officially ended with the discontinuation of the PlayStation (known in its re-engineered form as the "PSOne") in March 2006, a few months after the launch of the seventh generation.

Some features that distinguished fifth generation consoles from fourth generation consoles include:

- 3D polygon graphics with texture mapping

- Optical disc (CD-ROM) game storage, allowing much larger storage space (up to 650 MB) than ROM cartridges

- CD quality audio recordings (music and speech), PCM audio with 16-bit depth and 44.1 kHz sampling rate

- Wide adoption of full motion video, displaying pre-rendered computer animation or live action footage

- Display resolution from 480i to 576i

- Color depth up to 16,777,216 colors (24-bit true color)

- 3D graphical capabilities such as lighting, Gouraud shading, anti-aliasing and texture filtering

### 4.1.1 History

**Transition to 3D**

The 32-bit/64-bit era is most noted for the rise of fully 3D polygon games. While there were games prior that had used three-dimensional polygon environments, such as *Virtua Racing* and *Virtua Fighter* in the arcades and *Star Fox* on the SNES, it was in this era that many game designers began to move traditionally 2D and pseudo-3D genres into 3D on video game consoles. Early efforts from then-industry leaders Sega and Nintendo saw the introduction of the Sega 32X and Super FX which provided rudimentary 3D capabilities to the 16-bit Mega Drive/Genesis and SNES. Prime examples of this trend include *Virtua Fighter* and *Tomb Raider* (later ported to the PlayStation) on the Saturn, *Tekken* and *Crash Bandicoot* on the PlayStation, and *Super Mario 64* on the N64. Their 3D environments were widely marketed and they steered the industry's focus away from side-scrolling and rail-style titles, as well as opening doors to more complex games and genres. 3D became the main focus in this era as well as a slow decline of cartridges in favor of CDs, due to the ability to produce games less expensively and the media's high storage capabilities.

## CD vs. cartridge

See also: ROM cartridge

After allowing Sony to develop a CD-based prototype console for them and a similar failed partnership with Philips,[2] Nintendo decided to make the Nintendo 64 a cartridge-based system like its predecessors. Publicly, Nintendo defended this decision on the grounds that it would give games shorter load times than a compact disc (and would decrease piracy). However, it also had the dubious benefit of allowing Nintendo to charge higher licensing fees, as cartridge production was considerably more expensive than CD production. Many third-party developers like EA Sports viewed this as an underhanded attempt to raise more money for Nintendo and many of them became more reluctant to release games on the N64.

Nintendo's decision to use a cartridge based system sparked a small scale war amongst gamers as to which was better. The chief advantages of the CD-ROM format were (1)larger storage capacity, allowing for a much greater amount of game content, and (2)considerably lower manufacturing costs, making them considerably more profitable for game publishers. Its disadvantages compared to cartridge were (1)considerable load times, (2)their inability to load data "on the fly", making them reliant on the console RAM, and (3)the greater manufacturing costs of CD-ROM drives compared to cartridge slots, resulting in generally higher retail prices for CD-based consoles.[3] A Nintendo magazine ad placed a Space Shuttle (cartridge) next to a snail (a CD) and dared consumers to decide "which one was better".

Almost every other contemporary system used the new CD-ROM technology (the Nintendo 64 was the last major home video game console to use cartridges). Consequent to the storage and cost advantages of the CD-ROM format, many game developers shifted their support away from the Nintendo 64 to the PlayStation. One of the most influential game franchises to change consoles during this era was the *Final Fantasy* series, beginning with *Final Fantasy VII*, which was originally being developed for the N64 but due to storage capacity issues was shifted to and released on the PlayStation; prior Final Fantasy games had all been published on Nintendo consoles – either the Nintendo Entertainment System or Super Nintendo, with the only other entries being on computers like the MSX.

## Overview of the fifth generation consoles

There was much confusion over which console was superior to the others. Adding to the uncertainty was the fact that there were more competing consoles in this era than at any other time since the North American video game crash of 1983, with video game magazines frequently predicting a second crash due to the similar deluge of new consoles.[4] Also, console makers routinely boasted theoretical maximum limits of each system's 3D polygon rendering without accounting for real world in-game performance.

The FM Towns Marty is considered the world's first 32-bit console (predating the Amiga CD32 and 3DO), being released in 1991 by Japanese electronic company Fujitsu. Never released outside of Japan, it was largely marketed as a console version of the FM Towns home computer, being compatible with games developed for the FM Towns. It failed to make an impact in the marketplace due to its expense relative to other consoles and inability to compete with home computers.[5] While using a 32-bit word length, however, the Intel 80386SX CPU only supports 16-bit bus addressing (similar to the Motorola 68000 in 1985's Amiga 1000) and a maximum of 24-bit RAM addressing.

Despite massive third party support and an unprecedented amount of hype for a first-time entrant into the industry, the 3DO Interactive Multiplayer's $700 price tag hindered its success,[6] selling 2 million units world wide.

The Amiga CD32 was sold in Europe, Australia and Canada, but never in the United States due to Commodore's bankruptcy.[7]

The Sega 32X, an add-on console for the Mega Drive/Genesis and Sega CD, was launched a short time apart from the Sega Saturn. The Sega Neptune was also announced as a standalone version of the 32X/CD, but ultimately canceled. Sega failed to deliver a steady flow of games for the 32X platform. More importantly, with the Saturn and PlayStation already on the horizon, most gamers preferred to save up their money rather than spend it on a console that was doomed to become obsolete in just a few months.[8]

The Sega Saturn was released as Sega's entry into the 32-bit console market.[2] It sold 9.5[6] million units worldwide. It became Sega's most successful console in Japan, but it was not the overseas commercial success that the Master System and Mega Drive had been and lagged in third place overall.

The Atari Jaguar was released in 1993 as the world's first 64-bit system. However, sales at launch were well below the incumbent fourth generation consoles, and a small games library rooted in a shortage of third party support made it impossible for the Jaguar to catch up, selling below 300,000 units. The system's 64-bit nature was also questioned by many. The 32-bit Atari Panther, set to be released in 1991, was canceled due to unexpectedly rapid progress in developing the Jaguar.[9]

The Atari Jaguar CD, an add-on console for the Jaguar, was

released in 1995. It was produced in limited quantities due to the low install base of the system.

The PlayStation was the most successful console of this generation, with attention given by 3rd party developers enabling it to achieve market dominance, becoming the first home console to ship 100 million units worldwide.

Because of many delays in the release of the Nintendo 64, in 1995 Nintendo released the Virtual Boy, a supposedly portable system capable of displaying true 3D graphics, albeit in monochromatic red and black. However, in practice it is not functionally portable, though it was at first marketed as such, and because of the nature of its graphical capabilities, the system can cause headaches and eye strain. It was discontinued within a year, with less than 25 games ever released for it.

The Nintendo 64, originally announced as the "Ultra 64", was released in 1996. The system's delays and use of the cartridge format while all of its competitors used CDs made it an unpopular platform among third party developers. However, a number of wildly popular 1st party titles allowed the Nintendo 64 to maintain strong sales in the United States, though it still remained a distant second to the PlayStation.

NEC, creator of the TurboGrafx-16, TurboDuo, Coregrafx, and SuperGrafx, also entered the market with the PC-FX in 1994. The system had a 32-bit processor, 16-bit stereo sound, a 16,777,000 color palette and featured the highest quality full motion video of any console on the market at the time. The PC-FX broke away from traditional console design by being a tower system that allowed for numerous expansion points including a connection for NEC's PC-9800 series of computers. Despite its impressive specs, it was marketed as the ultimate side-scrolling console and could not match the sales of the 3D systems currently on the market. They had also lost developer support by their past partners, including Hudson Soft, who contributed only one game.

**Results of the fifth generation**

By the end of the 1995 Christmas shopping season, the fifth generation has come down to a struggle between the Sony PlayStation, Sega Saturn, 3DO Interactive Multiplayer, and the upcoming Nintendo 64. The FM Towns Marty and Amiga CD32 had already been discontinued; the Atari Jaguar and Sega 32X were still on the market but were considered a lost cause by industry analysts; the Neo Geo CD had proven to appeal only to a niche market; and industry analysts had already determined that the just-launched Apple Bandai Pippin was too expensive to make any impact in the market.[10]

After the dust settled in the fifth generation console wars, several companies saw their outlooks change drastically. Atari, which was not able to recover its losses, ended up being purchased by JT Storage and stopped making game hardware until the brand was revived for the Atari Flashback in 2004. Sega's loss of consumer confidence (coupled with its previous console failures) along with their financial difficulties, set the company up for a similar fate in the next round of console wars.

The Sega Saturn, ironically suffered from poor marketing and comparatively limited third-party support outside of Japan.[2] Sega's decision to use dual processors was roundly criticized, as this made difficult to efficiently develop for the console. Regardless of their reasons for including it, only Sega's first-party developers were ever able to use the second CPU effectively. The Saturn was far more difficult than the PlayStation to program for.

Sega was also hurt by a surprise four-month-early U.S. launch of their console. Third party developers, who had been planning for the originally scheduled launch, could not provide launch titles and were angered by the move. Retailers were caught unprepared, resulting in distribution problems. Some retailers, such as the now defunct KB Toys, were so furious that they refused to stock the Saturn thereafter.[11]

Due to numerous delays, the Nintendo 64 was released one year later than its competitors. By the time it was finally launched in 1996, Sony had already established its dominance, the Saturn was starting to struggle, and the Jaguar and 3DO had been discontinued. Its use of cartridge media rather than compact discs alienated some developers and publishers due to the space limits, the relatively high cost involved, and a considerably longer production time. In addition, the initially high suggested retail price of the console may have driven potential customers away, and some early adopters of the system who had paid the initial cost may have been angered by Nintendo's decision to reduce the cost of the system US$50 six months after its release. However, the Nintendo 64 was popular in North America, mostly the U.S, selling 20.63 million units in the region (more than half of its worldwide sales of 32.93 million units), and is home to highly successful games such as *The Legend of Zelda: Ocarina of Time*, *Super Mario 64*, *GoldenEye 007*, *Banjo-Kazooie*, and *Super Smash Bros.*. Still, while the Nintendo 64 sold far more units than the Sega Saturn, Atari Jaguar, and Panasonic/Goldstar/Sanyo 3DO, it failed to surpass the PlayStation, which dominated the market.

By 1997, 40% to 60% of American homes played on video game consoles. 30% to 40% of these homes owned a console, while an additional 10% to 20% rented or shared a console.[12]

## 4.1.2 Home systems

Technical Comparison[13]

### Other consoles

### Mass market

**These consoles were created for the mass market, like the 5 consoles listed above. These, however, are less notable, never saw a worldwide release, and/or have sold particularly poorly, and are therefore listed as 'Other'.**

- FM Towns Marty, created by Fujitsu. Released on February 20, 1993.

- Amiga CD32, created by Commodore. Released on September 17, 1993.

- CPS Changer, created by Capcom. Released in 1994.

- PC-FX, created by NEC. Released on December 23, 1994.

- Apple Bandai Pippin, created by Apple and Bandai. Released on March 28, 1995.

### Non-mass-market systems

- Playdia, created by Bandai. A console consisting of quiz games. Released in Japan on September 23, 1994, for ¥24,800.

- Casio Loopy, created by Casio. Released in October 1995 in Japan, targeted at female gamers.

### Add-ons and remakes

- Atari Jaguar CD
  Released in September 1995[1]

- Nintendo 64DD
  Released in Japan in 1999

1. ^ "Atari Jaguar CD system pounces onto multimedia marketplace". *Business Wire*. September 21, 1995. Retrieved 2011-05-07.

### Worldwide sales standings

See also: List of best-selling game consoles

From 1996 to 1999 (when the PlayStation, N64 and Saturn were the major 5th-generation consoles still on the market)

Sony managed a 47% market share of the worldwide market, followed by Nintendo with 28% (with a percentage of that figure from the 16-bit SNES), while Sega was third with 23% (with a percentage of that from the Dreamcast).[46]

Production of the Sega Saturn was discontinued in 1999. Its demise being accelerated by rumors that work on its successor was underway; these rumors hurt the systems' sales in the west as early as 1997. The N64 was succeeded by the GameCube in 2001, but continued its production until 2004; however, PlayStation production was not ceased as it was redesigned as the PSone, further extending the life of the console around the release of the follow-up PlayStation 2. The PlayStation console production was discontinued in 2006, shortly after the Xbox 360 was released.

## 4.1.3 Handheld systems

See also: List of handheld game consoles and Comparison of handheld game consoles

- Sega Nomad
  Released in 1995[1]

- R-Zone
  Released in 1995

- Virtual Boy
  1995–1996[2]

- Game Boy Pocket
  Released in 1996

- Game.com
  Released in 1997

- Game Boy Light
  Released in April, 1998

- Game Boy Color
  Released in November, 1998[3]

- Neo Geo Pocket
  Released in 1998

- Neo Geo Pocket Color (1998) Created by SNK

- WonderSwan (1999) Released in Japan only

- WonderSwan Color (2000) Released in Japan only

1. ^ *Retro Gamer* staff. "Retroinspection: Sega Nomad" . *Retro Gamer* (Imagine Publishing) (69): 46–53.

2. ^ Cite error: The named reference gamepro2 was invoked but never defined (see the help page).

3. ^ "Nintendo Adds Color to Its "Rainbow" of Products With New Game Boy Color Titles" . *Business Wire.* October 12, 1998. Retrieved 2011-05-07.

### 4.1.4 Software

Main article: List of console game franchises

**Milestone titles**

- *Castlevania: Symphony of the Night* (PlayStation, Saturn) by Konami Computer Entertainment Tokyo and Konami is considered one of the best PlayStation games available, and a strong argument for the relevance of 2D games in an increasingly 3D market.*[47]*[48]*[49]

- *Crash Bandicoot* (PlayStation) by Naughty Dog and Sony Computer Entertainment (SCE) would go on to become Sony's unofficial mascot in contrast with Nintendo's *Mario* and Sega's *Sonic the Hedgehog*. The game featured a marsupial bandicoot named Crash and would prove to be one of the PlayStation's most successful titles.

- *Dragon Warrior VII* (PlayStation) by Heartbeat, ArtePiazza, and Enix was the number one best-selling title on the PlayStation in Japan, released in 2000. The game was the first main installment of Japan's national RPG series released in 5 years.

- *FIFA International Soccer* (3DO) by EA Sports and Electronic Arts has been described as one of the most influential sports games ever made.*[50]

- *Final Fantasy VII* (PlayStation) by Square Product Development Division 1 and Squaresoft, is one of the PlayStation's most popular titles. It was the first game in the Final Fantasy series to make use of full motion videos (FMVs) and opened the door to the mainstream US market for Japanese-origin RPGs by SquareSoft. *Final Fantasy* became one of the biggest franchises in video gaming, with FFVII in particular having several spin-offs known as Compilation of Final Fantasy VII, including two sequels (a movie, and an action adventure game) and a prequel.

- *Gran Turismo* (PlayStation) by Polyphony Digital and SCE broke away from the mold of traditional arcade style racing games by offering realistic physics and handling as well as a plethora of licensed vehicles.

- *GoldenEye 007* (N64) by Rare Ltd. and Nintendo is a critically acclaimed game that helped make the first-person shooter a potential popular genre on consoles.

- *Metal Gear Solid* (PlayStation) by Konami Computer Entertainment Japan and Konami received critical acclaim for its involved storyline, believable voice acting, and cinematic presentation, and is considered one of the best games of all time. The series remains a best seller for the PlayStation along with the series branching off to Xbox and other Nintendo consoles after many successes.

- *The Need for Speed* (3DO, PlayStation, Saturn) by Pioneer Productions and Electronic Arts shot well ahead of prior racing simulators in graphics and realism, and spawned a number of sequels.

- *Nights into Dreams...* (Saturn) by Sonic Team and Sega was bundled with the Saturn's analog controller, which was almost essential to the gameplay. With its innovative gameplay and graphics, *Nights*, an exclusive title, aided in the selling of a number of Saturns.

- *Panzer Dragoon Saga* (Saturn) by Team Andromeda and Sega is the highest-rated Saturn title on Game Rankings with a score of 92.87%,*[51] and has been cited as one of the greatest games ever made.*[49]*[52]*[53]

- *PaRappa the Rapper* (PlayStation) by NanaOn-Sha and SCE, although only a modest success at its time of release, was highly influential in creating the music video game genre, which would grow in popularity throughout the fifth and sixth generations, thanks in large part to the popular Dance Dance Revolution.

- *Perfect Dark* (N64) by Rare Ltd. and Nintendo was critically acclaimed, building on what made *GoldenEye 007* successful, making use of the Nintendo 64 expansion pack. It received a 97/100 from metacritic and a 9.8 from IGN. The game was remade for its tenth anniversary in 2010 to positive reviews.*[54]*[55]*[56]

- *Pokémon Red and Blue* (GB) by Game Freak for Nintendo, led to a massive success in both video game sales and licensed merchandise. In addition to establishing a wildly popular franchise, Pokémon arguably helped extend the life of the handheld Game Boy system.

- *Resident Evil* (PlayStation, Saturn) by Capcom and *Silent Hill* (PlayStation) by Konami Computer Entertainment Tokyo and Konami helped popularize the survival horror genre on consoles. This genre continued to grow in the sixth generation of video games, and *Silent Hill* and *Resident Evil* went on to produce many successful sequels. Both have since been adapted for films.

- *Sega Rally Championship* (Arcade, Saturn) by Sega AM5 and Sega was the first rally racing game,*[57] broke new ground by incorporating different surfaces with different friction properties,*[58]*[59] and has been cited as one of the greatest racing games ever made.*[58]*[60]

- *Super Mario 64* (N64) by Nintendo Entertainment Analysis & Development (Nintendo EAD) and Nintendo is considered to be one of the greatest games of all time, particularly for its use of a dynamic camera system, the implementation of its 360-degree analog control, and open world design.*[61] Super Mario 64 is one of the best selling home console games of the era, selling 11.62 million copies worldwide.

- *Super Smash Bros.* (Nintendo 64) was a breakthrough IP for Nintendo, featuring characters from Nintendo owned franchises fighting in a party styled game. Super Smash Bros. has since been succeeded by 4 additional titles in the series.

- *Spyro the Dragon* (PlayStation) by Insomniac Games and SCE was another defining title of the PlayStation. The game was released in 1998 to much critical acclaim, spawning a best-selling trilogy.

- *Star Fox 64* (N64) by Nintendo EAD and Nintendo is the first Nintendo 64 game to use the Nintendo 64 Rumble Pak, which bundled with the game. It was a success and sold 3 million copies worldwide.

- *Tekken 3* (PlayStation) is considered not only to be the greatest installment of the *Tekken* series, but remains as one of the greatest fighting games of all time according to PlayStation Magazine.*[62] It has a Metacritic score of 96, and is the 12th highest rated game ever according to Game Rankings.*[63] Its predecessor achieved similar feats until its succession,*[64] and the first game in the franchise was the first PlayStation game to sell over a million units.*[65]*[66]

- *Tomb Raider* (PlayStation, Saturn) by Core Design and Eidos Interactive popularized many elements seen in later video games and spawned several very successful sequels. The main character, Lara Croft, was named the most recognizable female video game character by Guinness World Records.*[67] Two film adaptations have been released, the first of which, titled *Lara Croft: Tomb Raider* (2001), is the highest grossing video game to film adaptation as of 2014.*[67]

- *Tony Hawk's Pro Skater 2* (N64, PlayStation) by Neversoft and Activision garnered widespread critical acclaim and has been cited as one of the greatest games ever made.*[49]

- *Virtua Cop* (Arcade, Saturn) by Sega AM2 and Sega introduced the use of 3D polygons to the light-gun shooter genre,*[68] paving the way for future light gun shooters like Namco's *Time Crisis* and Sega's *The House of the Dead*, and was a major influence on *GoldenEye 007*.*[69]

- *Virtua Fighter* (Arcade, Saturn) by Sega AM2 and Sega created the 3D fighting game genre.*[70] The console port, which was nearly identical to the arcade game, sold at a nearly 1:1 ratio with the Saturn hardware at launch.*[71] The original arcade version also had a major influence on the PlayStation becoming a 3D focused console.*[72]

- *The Legend of Zelda: Ocarina of Time* (N64) by Nintendo EAD and Nintendo is one of the most critically acclaimed games of all time and often listed as one of the greatest video games of all time.*[49]*[73]*[74]*[75]*[76]*[77]*[78] It transferred the playing mechanics of the previous 2D Zelda adventures to a 3D environment, with a 3rd person perspective that could switch to 1st person view. It also featured mini-games, such as fishing and horseback riding, and gave birth to the Z targeting system, which would become a mainstay in the series battles.

### 4.1.5   See also

- Commodore International

- The 3DO Company

- Playdia

### 4.1.6   References

[1] "System List". GameFAQs. Retrieved 2012-08-15.

[2] Christopher Dring, 2013-07-11, A Tale of Two E3s - Xbox vs Sony vs Sega, MCV

[3] "The Format of the Future: CD-ROM or Cartridge?". *GamePro* (59) (IDG). June 1994. p. 8.

[4] Carpenter, Danyon (July 1994). "The Flood Waters Are Rising...". *Electronic Gaming Monthly* (60) (EGM Media, LLC). p. 6.

[5] "FM Towns Marty/FM Towns Marty 2 Console Information". Consoledatabase.com. Archived from the original on 2005-05-07. Retrieved 2009-08-17.

[6] Blake Snow (2007-05-04). "The 10 Worst-Selling Consoles of All Time". GamePro.com. p. 1. Archived from the original on 2008-09-05. Retrieved 2007-11-25.

[7] Perelman, M: "Steal This Idea", page 60. Palgrave Macmillan, 2004

[8] "32X/Project Mars: Anatomy of a Failure". goodcowfilms.com. Retrieved 2007-06-22.

[9] Atari Jaguar History, AtariAge.

[10] "1996". *Electronic Gaming Monthly* (Ziff Davis) (78): 18–20. January 1996.

[11] Helgeson, Matt. "Top 10 Embarrassing E3 Moments", *Game Informer*(208): 40–41.

[12] http://webcache.googleusercontent.com/search?q=cache: 2W7jV8xhO_QJ:www.economics.rpi.edu/public_ html/ruiz/EGDFall2013/readings/From%2520Barbie% 2520to%2520Mortal%2520Combat.doc

[13] http://www.angelfire.com/electronic2/mariotan/

[14] "Will the Release of the PSX Ignite Gamers' Interests?". *Electronic Gaming Monthly* (Ziff Davis) (74): 26–27. September 1995.

[15] "At the Deadline". *GamePro* (IDG) (85): 174. October 1995.

[16] "Tidbits...". *Electronic Gaming Monthly* (Ziff Davis) (76): 19. November 1995.

[17] "Japan Platinum Game Chart". The Magic Box. Retrieved 2007-11-25.

[18] "Gran Turismo Series Shipment Exceeds 50 Million Units Worldwide" (Press release). Sony Computer Entertainment. 2008-05-09. Archived from the original on 2013-06-18. Retrieved 2008-06-03.

[19] ""Gran Turismo" Series Software Title List". Polyphony Digital. April 2008. Retrieved 2008-06-03.

[20] "Mario sales data". GameCubicle.com. Retrieved 2007-11-25.

[21] "All Time Top 20 Best Selling Games". 2003-05-21. Archived from the original on 2006-02-21. Retrieved 2007-11-25.

[22] http://www.drolez.com/retro/

[23] http://www.segatech.com/technical/saturnspecs/

[24] http://www.dcshooters.co.uk/sega/saturn/saturn.php

[25] "Saturn Overview Manual" (PDF). Sega of America. 1994-06-06. Retrieved 2014-04-25.

[26] http://www.futuretech.blinkenlights.nl/sgi2.html

[27] http://koti.kapsi.fi/~{}antime/sega/files/ ST-013-R3-061694.pdf

[28] http://koti.kapsi.fi/~{}antime/sega/files/ ST-058-R2-060194.pdf

[29] http://psx.rules.org/gpu.txt

[30] http://www.consoledatabase.com/faq/segasaturn/ segasaturnfaq.txt

[31] http://koti.kapsi.fi/~{}antime/sega/files/ ST-077-R2-052594.pdf

[32] http://www.gamepilgrimage.com/NFSComp.htm

[33] http://www.hillsoftware.com/files/atari/jaguar/jag_v8.pdf

[34] http://www.ataritimes.com/index.php?page=Atari% 20Jaguar

[35] https://kris-genthe.squarespace.com/config

[36] http://www.system16.com/hardware.php?id=711

[37] http://www.gamepilgrimage.com/SATPScompare.htm

[38] http://www.sega-saturn.com/saturn/other/satspecs.htm

[39] http://www.8-bitcentral.com/images/sony/playstation/ boxBack.jpg

[40] https://archive.org/stream/nextgen-issue-001/Next_ Generation_Issue_001_January_1995#page/n47/mode/ 2up/

[41] http://n64.icequake.net/mirror/www.white-tower.demon. co.uk/n64/

[42] "PlayStation Cumulative Production Shipments of Hardware". Sony Computer Entertainment Inc. Archived from the original on 2011-05-24. Retrieved 2008-03-22.

[43] "05 Nintendo Annual Report - Nintendo Co., Ltd." (PDF). Nintendo Co., Ltd. 2005-05-26. p. 33. Retrieved 2007-11-25.

[44] Greg Orlando (2007-05-15). "Console Portraits: A 40-Year Pictorial History of Gaming". *Wired News.* Condé Nast Publications. Retrieved 2008-03-23.

[45] Blake Snow (2007-05-04). "The 10 Worst-Selling Consoles of All Time". GamePro.com. p. 2. Archived from the original on 2008-09-05. Retrieved 2007-11-25.

[46] "New Versatility in Video Game Consoles Helps Boost Sales". In-Stat (NPD Group). January 23, 2001. Archived from the original on 2005-02-19. Retrieved 31 January 2012.

[47] Varanini, Giancarlo. "GameSpot Greatest Games of All Time: Castlevania: Symphony of the Night". *GameSpot.com.* Archived from the original on 2010-06-16. Retrieved 2014-01-18.

[48] "Top 100 games of All Time (2005)". ign.com. Retrieved 2014-01-18.

[49] Cork, Jeff (2009-11-16). "Game Informer's Top 100 Games of All Time (Circa Issue 100)". *Game Informer.* Retrieved 2014-01-18.

[50]  Kent, Steven L. (2006-10-09). "SOMETIMES THE BEST" . Sad Sam's Place. Retrieved 2014-02-02.

[51]  http://www.gamerankings.com/saturn/ 198258-panzer-dragoon-saga/index.html

[52]  "IGN Top 100 Games 2007" . IGN.com. Retrieved November 24, 2008.

[53]  Top 100 Games of All Time: No.22, G4.

[54]  "Perfect Dark for Nintendo 64 - Reviews, Ratings, Credits, and More" . Metacritic. 2000-05-22. Retrieved 2013-06-30.

[55]  "Perfect Dark - Nintendo 64". IGN. 2000-04-08. Retrieved 2013-06-30.

[56]  "Perfect Dark XBLA Review" . IGN. Retrieved 2013-06-30.

[57]  "The Making Of: Sega Rally Championship 1995" . Edge. Future plc. 2 October 2009. Retrieved January 18, 2014.

[58]  Guinness World Records: Gamer's Edition 2009, page 103.

[59]  Edge Staff, The Making Of: Colin McRae Rally, Edge, February 5, 2010: "The basic premise for the game was based around the car handling in Sega Rally," confirms Guy Wilday, producer of the first four CMR games. "Everyone who played it loved the way the cars behaved on the different surfaces, especially the fact that you could slide the car realistically on the loose gravel. The car handling remains excellent to this day and it's still an arcade machine I enjoy playing, given the chance."

[60]  "Top 25 Racing Games... Ever! Part 2" . Retro Gamer. 21 September 2009. pp. 5–6. Retrieved 2014-01-18.

[61]  "The Essential 50 Part 36: Super Mario 64" . 1UP.com. Retrieved February 13, 2014.

[62]  PlayStation: The Official Magazine asserts in its January 2009 issue that Tekken 3 "is still widely considered one of the finest fighting games of all time." See "Tekken 6: A History of Violence," PlayStation: The Official Magazine (January 2009): 46.

[63]  http://www.gamerankings.com/browse.html

[64]  Staff (September 1997). "Top 25 PlayStation Games of All Time" . PSM 1 (1): 34.

[65]  http://www.absolute-playstation.com/api_faqs/faq20.htm

[66]  http://www.dualshockers.com/2010/08/02/ then-and-now-the-history-of-tekken/

[67]  "Record-Breaking Lara Croft Battles her Way Into New Guinness World Records" , MCV. January 21, 2010.

[68]  Virtua Cop, IGN, July 7, 2004.

[69]  Martin Hollis (2004-09-02). "The Making of GoldenEye 007". Zoonami. Archived from the original on 2011-07-18. Retrieved 2014-01-18.

[70]  Leone, Matt, Essential 50: Virtua Fighter, 1UP.

[71]  Kent, Steven L. (2001). The Ultimate History of Video Games: The Story Behind the Craze that Touched our Lives and Changed the World. Roseville, California: Prima Publishing. p. 502. ISBN 0-7615-3643-4.

[72]  Feit, Daniel (2012-09-05). "How Virtua Fighter Saved PlayStation's Bacon" . Wired. Retrieved 2014-10-09. Ryoji Akagawa: If it wasn't for Virtua Fighter, the PlayStation probably would have had a completely different hardware concept.

[73]  "The Legend of Zelda: Ocarina of Time reviews" . Metacritic. Retrieved 2008-11-26.

[74]  "IGN Top 100 Games, #001-010 (2005)". IGN. Retrieved 2008-11-26.

[75]  "IGN Top 100 Games, #4 (2007)". IGN. Retrieved 2008-11-26.

[76]  "NP Top 200", Nintendo Power 200: 58–66, February 2006.

[77]  "The Greatest 200 Games of Their Time" , Electronic Gaming Monthly 200: February 2006.

[78]  "All-Time Best Rankings" . GameRankings. Retrieved 2008-11-26.

## 4.2  List of Sega arcade system boards

The following is a **list of arcade system boards** released by **Sega**. For games running on these system boards, see List of Sega arcade games.

### 4.2.1  Sega Blockade

Sega Blockade was Sega's first unified arcade system board, released in 1976 for Blockade and then used for other similar games in 1976 and 1977.*[1]

- CPU: Intel 8080A @ 2 MHz*[2] (8-bit instructions @ 0.29 MIPS)*[3]

- Sound chip: Custom*[1] (mono)

- Colors: 2*[2] (monochrome)

- Display resolution: 256×224 pixels*[2]

- Refresh rate: 60 Hz*[2]

### 4.2.2 Sega VIC Dual

Sega released the *Sega VIC Dual* arcade system board in 1977 as one of the first systems to use the Zilog Z80 microprocessor. Some of the games on the system include *Depthcharge* (1977), *Frogs* (1978), *Heiankyo Alien* (1979), *Head On* (1979), *Carnival* (1980), and *Samurai* (1980).*[4]

- CPU: Zilog Z80 @ 1.934 MHz*[5] (8-bit & 16-bit instructions @ 0.28 MIPS),*[6] or Intel 8080 @ 1.934 MHz*[4] (8-bit instructions @ 0.28 MIPS)*[3]

- Sound board: Custom (mono)

- Color model: Monochrome (1977) or RGB (1979)*[5]

- Color palette: 64*[7]

- Colors on screen: 2 (1977) or 8 (1979)*[8]

- Display resolution: 256×224*[4] to 328×262*[5]

- Refresh rate: 60 Hz*[5]

### 4.2.3 Sega Z80

*Sega Z80* was an arcade system board that is named after the Zilog Z80 processor it uses as its main CPU. It released in 1980, with games such as *Moon Cresta*,*[9] using a modified version of the Namco Galaxian system board.*[10] In 1981, *Jump Bug* added parallax scrolling*[11] and replaced the sound chip.*[12] In 1982, *Super Locomotive* replaced the Namco Galaxian hardware with more advanced custom Sega hardware,*[9] including sound and graphics chips that would later be used in the System 1/2/16 and Sega Space Harrier boards.

#### Specifications

- Main CPU: Zilog Z80 @ 3.072 MHz*[9] (8-bit & 16-bit instructions @ 0.45 MIPS*[6])

- Sound hardware: Namco Galaxian sound hardware (one programmable 4/8-bit waveform channel, three 4-bit square wave channels, two 17-bit noise channels, one modulated noise pulse channel)*[10]

- GPU chipset: Namco Galaxian video hardware*[10]*[13]

- Display resolution: 256×224*[14] to 384×264*[15] (horizontal), 224×256 to 264×384 (vertical)

- Refresh rate: 60.60606 Hz (V-sync)*[10]

- Color model: RGB*[16]

- Color palette table: 224 (PROM)*[13]

- Colors on screen: 32 (palette RAM)*[10]

- Background planes:
  - Tilemap plane: 8×8 tile sizes,*[10] scrolling*[13]
  - Bitmap plane: Star generator, scrolling*[13]

- Sprite capabilities: 8×8 to 16×16 sizes, 4 colors per sprite,*[10] 15 sprites per scanline, 240 sprite pixels per scanline, sprite flipping,*[13] sprite animation*[17]

*Jump Bug* added the following upgrades in 1981:

- Sound chip: General Instrument AY-3-8910 @ 1.78975 MHz*[10]

- Background planes: Parallax scrolling*[11]

*Super Locomotive* included the following upgrades/modifications in 1982:

- Main CPU: Zilog Z80 @ 4 MHz*[9] (8-bit & 16-bit instructions @ 1.16 MIPS*[6])

- Audio CPU: Zilog Z80 @ 4 MHz*[9] (8-bit & 16-bit instructions @ 1.16 MIPS*[6])

- Sound chips: Sega SN76496*[9] (modified Texas Instruments SN76496*[18]) @ 4 MHz, Sega SN76496 @ 2 MHz*[9]

- GPU chipset: Sega 315-5011*[19] (sprite line comparator),*[20] Sega 315-5012*[19] (sprite generator)*[20]

- Display resolution: 248×224 to 256×256*[19]

- Refresh rate: 60 Hz*[19]

- Color palette table: 1568*[19]

- Colors on screen: 768*[19]*[21]

- Color per sprite: 16*[19]

- Sprite pixels: 4 MHz clock cycles (60 Hz refresh rate),*[19] 66,666 pixels per frame (256 scanlines), 260 sprite pixels per scanline, 32 sprites per scanline

- Tilemap planes: 2 background layers*[21]

*Bank Panic* included the following upgrades/modifications in 1984:

- Sound chips: 3× Sega SN76496 @ 3.86712 MHz*[9]

- Tilemap planes: 2 layers (foreground, background)*[22]

- Display resolution: 256×224*[14] to 330×256*[23]

### 4.2.4  Sega G80

*Sega G80* was an arcade system board released by Sega in 1981. The G80 was released in both raster and vector versions of the hardware.

**G80 specifications**

- Main CPU:[*][24]

  - Raster: Zilog Z80 @ 8 MHz (8-bit & 16-bit instructions @ 1.16 MIPS[*][6])
  - Vector: Zilog Z80 @ 3.86712 MHz (8-bit & 16-bit instructions @ 0.561 MIPS[*][6])

- Sound boards:

  - Sega USB (Universal Sound Board)[*][25]

    - MCU: Intel i8035[*][26] @ 3.12 MHz[*][27] (8-bit instructions @ 3.12 MIPS, 1 instruction per cycle)[*][28]
    - Sound chip: Sega Melody Generator[*][29] (programmable sound generator)

  - Speech Board[*][24] (optional)[*][27]

    - MCU: Intel i8035/i8039[*][24] @ 3.12 MHz[*][27] (8-bit instructions @ 3.12 MIPS)[*][28]
    - Speech synthesizer: General Instrument SP0250 (linear predictive coding)

  - Sega System 1 sound board (optional, used for *Sindbad Mystery* in 1983)[*][25]

    - Sound chips: Sega SN76496 @ 4 MHz, Sega SN76496 @ 2 MHz

- Raster graphics board: Sega Video I[*][25]

  - Raster display controller: Sega Raster Display Controller[*][24] @ 15.468 MHz[*][25]
  - Display resolution: 256×224[*][26] to 328×262 (horizontal),[*][25] 224×256 to 262×328 (vertical)[*][30]
  - Color palette table: 256 (8-bit RGB PROM)[*][24]
  - Colors on screen: 64 to 128 (palette RAM)[*][24][*][25]
  - Tilemap planes: 2 layers, horizontal and vertical scrolling,[*][30] 8×8 tiles, 4 colors per tile[*][24][*][25]
  - Sprite capabilities: 28 to 32 sprites per scanline, 224 to 256 sprite pixels per scanline, 4 colors per sprite, 8×8[*][24] to 16×16[*][26] sizes

- Vector display controller: Sega Display Controller[*][24]

  - Color depth: 64 (6-bit RGB)[*][31]

### 4.2.5  VCO Object

*VCO Object*,[*][32] also known as *Sega Z80-3D system*,[*][33] was released by Sega in 1981. It was the first system specifically designed for pseudo-3D sprite-scaling graphics, using analog scaling. It was used for the third-person racing video game *Turbo* (1981), the stereoscopic 3D shooter game *SubRoc-3D* (1982), and the third-person rail shooter *Buck Rogers: Planet of Zoom* (1982).[*][32] *SubRoc-3D* also introduced an active shutter 3D system, jointly developed by Sega with Matsushita (now Panasonic).[*][34]

**Specifications**

- CPU:[*][32]

  - *Turbo & SubRoc-3D*: Z80 @ 5 MHz (8-bit & 16-bit instructions @ 0.725 MIPS[*][6])
  - *Buck Rogers*: 2× Z80 @ 5 MHz (8-bit & 16-bit instructions @ 1.45 MIPS[*][6])

- Sound board: Sega Sound Board[*][35]

  - Sound chip: Custom
  - Audio output: Stereo[*][36]

- Display resolution:[*][37][*][38]

  - Standard resolution: 256×224 to 320×264 (horizontal), 224×256 to 264×320 (vertical)
  - Analog scaling resolution: 512×224 to 640×264 (horizontal), 224×512 to 264×640 (vertical)

- Refresh rate: 60 Hz (V-sync)[*][37]

- Frame rate: 30 frames per second (*SubRoc-3D*),[*][34] or 60 frames per second (*Turbo, Buck Rogers*)[*][37]

- Color palette: 832 (*Turbo*), or 768 (*SubRoc-3D*), or 1536 (*Buck Rogers*)[*][37]

- Colors on screen: 256 (*Turbo, SubRoc-3D*), or 1024 (*Buck Rogers*)[*][37]

- Background planes:

  - Tilemap layer: 8×8 pixel tiles, 4 colors per tile, scrolling, tile flipping[*][37]
  - Bitmap layer[*][33]

- Sprite capabilities: Pseudo-3D sprite-scaling (analog scaling), line buffer,[33] 64 sprites on screen,[37] 16 sprites per scanline,[33] 4[37] to 8[32] colors per sprite

    - Sprite pixels: 4.992 MHz (standard) to 9.984 MHz (scaling) pixel clock,[37][38] 83,200 (standard) to 166,400 (scaling) pixels per frame, 315 (standard) to 630 (scaling) sprite pixels per scanline

- *SubRoc-3D* capabilities: Stereoscopic 3D, active shutter 3D system[34]

### 4.2.6 Sega Zaxxon

The *Sega Zaxxon* hardware was released by Sega in 1982 as the first system dedicated to producing isometric graphics, first used for the isometric shooter *Zaxxon* (1982). It was also used for several other games, including the isometric platformer *Congo Bongo* (1983).

**Specifications**

- Main CPU: Zilog Z80 @ 3.04125 MHz[39] (8-bit & 16-bit instructions @ 0.441 MIPS[6])

- Sound board: Sega G80 Sound Board

- Graphics board: Sega Zaxxon-VIDEOII / 834-5167 Video Board[40]

- Display resolution: 256×224[39] to 384×264[40] pixels

- Color palette table: 512 (9-bit RGB PROM)[40]

- Colors on screen: 256 (palette RAM)[40]

- Tilemap planes:[41] 2 layers (foreground, background),[42] 8×8 tiles, 4 or 8 colors per tile,[40] tile flipping, vertical/horizontal/diagonal scrolling,[42] isometric perspective

- Sprites: 4 or 8 colors per sprite,[40] sprite flipping,[42] shadows

    - Sprite sizes: 8 and 32 heights,[40] widths of 8,[40] 16[42] and 32[40] pixels

    - Line buffer: 256 sprite pixels per scanline,[42] 8 (32-width) to 32 (8-width) sprites per scanline

*Congo Bongo* added the following specifications in 1983:

- Additional CPU: Zilog Z80 @ 2 MHz[39]

- Sound board: Sega 834-5168 Sound Board[40]

- Additional sound chips: 2× SN76496 @ 4 MHz

- Colors on screen: 512[40]

### 4.2.7 Sega Laserdisc

The *Sega Laserdisc* hardware was released by Sega in 1983 as the first system dedicated to producing laserdisc video games. The first game to use it was *Astron Belt* (1983) and the last to use it was the holographic game *Time Traveler* (1991).

**Specifications**

- CPU: Zilog Z80 @ 5 MHz[43] (8-bit & 16-bit instructions @ 0.725 MIPS)[6]

- JAMMA board: Sega PCB CN1[43]

- Audio board: Sega PCB CN2 (stereo output)[43]

- Laserdisc player: Hitachi-Sega VIP-9500SG[44]

    - Video resolution: 580×480 (580 dots, 480 lines), 525 scanlines (480 visible),[45] interlaced video

    - Refresh rate: 59.94 Hz[43]

    - Frame rate: 29.97 frames per second

    - Color depth: 16,777,216 (24-bit true color)

    - Audio: LaserDisc, stereo output

- Graphics overlay:[43]

    - Display resolution: 256×256 pixels, progressive scan

    - Refresh rate: 59.94 Hz

    - Color palette table: 512 (PROM)

    - Colors on screen: 256 (color RAM)

    - Tilemap plane:

        - Tile size: 8×8 pixels

        - Tilemap size: 32×32 (1024) tiles, 256×256 pixels

        - Colors per tile: 2

    - Sprite plane:

        - Sprite sizes: 8×8 to 256×8 pixels

        - Sprites on screen: 32 sprites per scanline, 256 sprite pixels per scanline

### 4.2.8   Sega System series

**Sega System 1**

*Sega System 1* was a type of arcade hardware used in various Sega arcade machines from 1983 until 1987. For most of its run it coexisted with *Sega System 2* (1985–1988) and as a result had many similar features (the only major difference being that System 2 had two separate circuit boards instead of one). In its four-year span it was used in some 20 different arcade games, including *Choplifter*, *Flicky*, *Pitfall II: Lost Caverns*, *Wonder Boy*, and *Wonder Boy in Monster Land*. System 2 is an updated version of the System 1.

**System 1 specifications**

- Board composition: Arcade components were contained on one circuit board

- Main CPU: Zilog Z80 @ 4 MHz[46] (8-bit & 16-bit instructions @ 0.58 MIPS)[6]

- Sound CPU: Zilog Z80 @ 4 MHz[46] (8-bit & 16-bit instructions @ 0.58 MIPS)[6]

- Sound chips: Sega SN76496 @ 4 MHz, Sega SN76496 @ 2 MHz

- GPU: Sega 315-5011[47] (sprite line comparator),[20] Sega 315-5012[47] (sprite generator),[20] 315-5049[47] (tilemap chip)[20]

- Display resolution: 256×224 to 640×260 pixels[47]

- Refresh rate: 60.0952 Hz (V-sync)[47]

- Colors on screen: 2048[47]

- Color palette: 4096[48]

- Tilemap planes:[48] 2 background layers[46] (1 static, 1 scrolling),[48] 8×8 tiles[47]

  - Tilemap sizes: 256×256 for both planes (System 1), or 512×512 for scrolling plane and 256×256 for fixed plane (System 2)[48]

- Sprites: Dual line buffers, double buffering, 32 sprites per scanline, 16 colors per sprite, sprite flipping,[48] hardware collision detection,[46][48] 8[47] to 256[48] width, 8[47] to 256[48] height

  - Sprite pixels: 10 MHz pixel clock cycles (60.0952 Hz refresh rate),[47] 166,402 pixels per frame (260 scanlines), 640 sprite pixels per scanline

**Sega System 16**

The *Sega System 16* is an early 16-bit arcade system board released by Sega in 1985.[49] Over its lifespan, roughly forty games were released on this hardware, making it one of Sega's most successful arcade platforms. It was produced in two variants, the System 16A and System 16B. Some games released using this hardware include: *Shinobi*, *Golden Axe*, *Altered Beast*, and *Dynamite Dux*.

In order to prevent piracy, as well as illegal bootleg games, many System 16 boards used an encryption system. A Hitachi FD1094 chip, containing the main CPU as well as the decryption key, was used in place of a regular CPU.

The System 16's pairing of a Motorola 68000 CPU and a Zilog Z80 coprocessor would prove to be a popular and durable arcade hardware configuration well into the 1990s. Capcom's CPS-1 and CPS-2 boards were built on a similar foundation, as was SNK's Neo Geo hardware. Sega would later use the 68000/Z80 combination to power its Genesis/Mega Drive home console.

**System 16 specifications**

- Main CPU: Hitachi FD1094 (Motorola 68000) @ 10 MHz[50][51][52] (16-bit & 32-bit instructions @ 1.75 MIPS)[6]

- Main MCU: Intel i8751 @ 8 MHz[50] (8-bit instructions @ 8 MIPS, 1 instruction per cycle)[53]

- Sound CPU: NEC uPD780C-1 (Zilog Z80 clone)[54] @ 4 MHz[50][51] (8-bit & 16-bit instructions @ 0.58 MIPS)[6]

- Sound MCU: Intel i8048 @ 6 MHz[54] (8-bit instructions @ 6 MIPS)[28]

- FM synthesis sound chip: Yamaha YM2151 @ 4 MHz (8 FM synthesis channels)

- PCM sound chip: NEC uPD7751[51] @ 6 MHz[54]

  - ADPCM channels: 3[54]

  - Audio bit depth: 8-bit[54]

- GPU chipset: 315-5011 sprite line comparator, 315-5012 sprite generator, 2× 315-5049 tilemap chips, 315-5107 & 315-5108 display timers, 315-5143 & 315-5144 sprite chips, 315-5149 video mixer[20]

  - Performance: 12.5874 MHz sprite line buffer render clock, 6.2937 MHz sprite line buffer scan/erase & pixel clock[54]

- Memory: 16kB + 2 kB (System 16A)[55]

- Display resolution: 320×224*[51] to 342×262*[56] (horizontal), 224×320 to 262×342 (vertical), progressive scan

- Color palette: 98,304*[57]

- Colors on screen: 4096 (unique colors)*[51] to 6144 (with shadow & highlight)*[56]

- Graphical planes: 1 sprite layer, 1 text layer, 2 tile layers*[51]*[52] (row & column scrolling,*[58] 8×8 tiles)*[56]

- Sprite capabilities: Dual line buffers, double buffering,*[58] 128 on-screen sprites,*[51]*[52] 800 sprite pixels (800.75 sprite processing ticks) per scanline, 100 sprites per scanline,*[54] 16 colors per sprite,*[58] 8*[56] to 256*[54]*[58] width, 8*[56] to 256*[54]*[58] height

**System 16B specifications**  System 16B included the following upgrades in 1986:

- Sound CPU: Zilog Z80 @ 5 MHz*[52] (8-bit & 16-bit instructions @ 0.725 MIPS)*[6]

- PCM sound chip: NEC uPD7759 ADPCM Decoder @ 640 kHz*[52]

  - ADPCM channels: 8*[59]
  - Audio bit depth: 9-bit*[59]*[60]
  - Other features: 8 kHz sampling rate, up to 128 KB audio ROM and 256 samples*[59]

- GPU chipset: 315-5196 sprite generator, 315-5197 tilemap generator, 315-5213 sprite chip,*[20] 315-5248 & 315-5250 math chips*[58]

- Sprite capabilities: Sprite-scaling*[52]

## Sega System 24

The *Sega System 24* was an arcade system board released by Sega in 1988. It was produced for coin-operated video arcade machines until 1996. Some games released using this hardware include: *Bonanza Bros.*, *Hot Rod*, and *Gain Ground*.

**Sega System 24 specifications**  The System 24 used two Motorola 68000 processors at 10 MHz. One was for input/output, while the other was used by the game. The board holds 1360 kB of RAM and 256 kB of ROM. It was the first Sega arcade system that required a medium resolution arcade monitor. The color palette is 4352 on screen selectable from 32,768,*[61] or with shadow & highlight, 16,384*[62] on screen selectable from 98,304.*[57] The system could support up to 2048 sprites on-screen at once.

Sound was driven by a YM2151 at 4 MHz; it was capable of delivering 8 channels of FM sound in addition to a DAC used for sound effects and sampling. Early System 24s loaded their program from floppy disks. Games could also use hardware ROM boards to store games. No matter which storage device was used, a special security chip was required for each game an operator wanted to play.*[61]

- CPU: Hitachi FD1094 @ 10 MHz*[63] & Motorola 68000 @ 10 MHz*[62] (16-bit & 32-bit instructions @ 3.5 MIPS)*[6]

- Sound chips: Yamaha YM2151 @ 4 MHz (8 FM synthesis channels), DAC (sound effects and speech synthesis)*[61]

  - Audio output: Stereo speakers, stereo headphones*[61]

- GPU chipset: Sega 315-5242 Color Encoder 315-5292 Tilemap Generator, 315-5293 Sprite Generator, 315-5294 Priority Mixer, 315-5295 Object Generator*[20]*[62]

  - Graphical capabilities: Sprite zoom, scrolling, row & column scrolling,*[64] parallax scrolling

- Memory:

  - RAM: 1360 KB
  - ROM: 256 KB

- Storage media: Floppy disk, ROM board

- Resolution: 496×384 to 656×424 pixels,*[62] progressive scan

- Color palette: 32,768 (unique colors)*[61] to 98,304 (with shadow & highlight)*[57]

- Colors on screen: 4352 (unique colors)*[61] to 16,384 (with shadow & highlight)*[62]

- Tilemaps:

  - Tilemap layers: 4*[20] (2 scrolling and 2 windowed)*[61]
  - Tile size: 8×8 pixels*[20]
  - Tiles per scrolling tilemap: 4096*[61]
  - Scrolling tilemap size: 512×512 pixels

- Sprites:

  - Colors per sprite: Up to 256*[65]

- Sprite size: 8 to 1024 pixels in width/height*[65]
- Sprite virtual space: 4096×4096 pixels*[65]
- Sprite pixels per line: 4096
- Sprites per line: 4 (1024×1024 pixels each) to 512 (8×8 pixels each)
- Sprites on screen: 2048 on screen*[61]
- Framebuffers: 2 framebuffers @ 512×384 pixels each, double buffering*[64]

### Sega System 18

The *Sega System 18* is an arcade system board released by Sega in 1989. System 18 had a very short run of games but most boards on this hardware were JAMMA standard. Most of these games also have the "suicide battery" as associated with Sega's System 16 hardware. It also contained the VDP used by the Sega Mega Drive console.*[20]

### System 18 specifications

- Main CPU: Motorola 68000 @ 10 MHz*[66] (16-bit & 32-bit instructions @ 1.75 MIPS)*[6]
- Sound CPU: Zilog Z80 @ 8 MHz*[66] (8-bit & 16-bit instructions @ 1.16 MIPS)*[6]
- Sound chip: 2 × Yamaha YM3438 @ 8 MHz + Ricoh RF5c68 @ 10 MHz (8-channel PCM chip, remarked as Sega Custom 315)
- Graphics chips: Sega System 16B chipset, Yamaha YM7101 VDP*[20]
- Display resolution: 320 × 224
- Color palette: 98,304*[57]
- Colors on screen: 4096 (unique colors)*[66] to 8384 (with shadow & highlight)*[67]
- Board composition: Main board + ROM board
- Graphical capabilities: 128 sprites on screen at one time, 4 tile layers, 1 text layer, 1 sprite layer with hardware sprite zooming, translucent shadows,*[66] sprites of any height and length, row & column scrolling*[58]

### 4.2.9   Kyugo

*Kyugo* is an arcade system board released in 1984, co-developed with Japanese company Kyugo.*[68] It was used for three Sega games: *Flashgal* and *Repulse* in 1985, and *Legend* in 1986.*[69] It was also used by several other companies from 1984 to 1987.*[68]

### Kyugo specifications

- CPU: 2× Zilog Z80 @ 4.608 MHz*[69] (8/16-bit instructions @ 1.34 MIPS*[6])
- Sound chips: 2× General Instrument AY-3-8910 @ 1.5 MHz*[69] (6 PSG channels)
- Display resolution: 288×239 to 512×256 (horizontal),*[68] 224×288*[69] to 256×512 (vertical)
- Refresh rate: 60 Hz*[68]
- Color palette: 4096 (12-bit RGB)*[68]
- Colors on screen: 256*[69] (8-bit RGB)
- Tilemap capabilities: 3 planes*[68] (foreground, background, text),*[70] 8×8 tiles, 4 colors per foreground tile, 8 colors per background tile, vertical scrolling, side-scrolling,*[68] tile flipping*[70]
- Sprite capabilities: 16×16 size, 8 colors per sprite,*[68] sprite flipping,*[70] 2 KB sprite RAM, 32 bytes per sprite,*[68] 64 sprites on screen
- Sprite pixels: 320 sprite pixels per scanline,*[70] 20 sprites per scanline

### 4.2.10   Super Scaler series

### Sega Space Harrier

*Sega Space Harrier*, also known as *Sega Hang-On*, was an early 16-bit system released in 1985, originally designed for the racing game *Hang-On* and third-person rail shooter *Space Harrier* (1985). It was also used for the racing game *Enduro Racer* (1986). This was the first in Sega's Super Scaler series of pseudo-3D arcade hardware. At the time of its release, this was the most powerful game system.*[71]

The pseudo-3D sprite/tile scaling in Sega's Super Scaler arcade games were handled in a similar manner to textures in later texture-mapped polygonal 3D games of the 1990s.*[72] Designed by Sega AM2's Yu Suzuki, he stated that his "designs were always 3D from the beginning. All the calculations in the system were 3D, even from Hang-On. I calculated the position, scale, and zoom rate in 3D and converted it backwards to 2D. So I was always thinking in 3D." *[73]

### Specifications

- Main CPU: Motorola MC68000 & Hitachi FD1094 (Motorola 68000)*[74] @ 10 MHz*[75] (16-bit & 32-bit instructions @ 3.5 MIPS)*[6]

- MCU: Intel i8751 @ 8 MHz (*Space Harrier*)[75] (8-bit instructions @ 8 MIPS)[53]

- Sound CPU: Z80 @ 4 MHz[75] (8-bit & 16-bit instructions @ 0.58 MIPS)[6]

- Sound board: Sega 834-5670[74]

- Sound chips:

  - FM synthesis chip: Yamaha YM2151 @ 4 MHz[75] (8 FM channels)

  - PCM chip: SegaPCM[75] (315-5218[76]) @ 4 MHz[74] (stereo output, 16 PCM channels, 12-bit audio,[77] 31.25 kHz sampling rate[75])

- GPU: Sega Super Scaler chipset

  - Graphics chips: 315-5011 sprite line comparator, 315-5012 sprite generator, 2× 315-5049 tilemap chips, 2x 315-5107 horizontal timing control, 315-5108 vertical timing control, 315-5122 timing chip[20]

  - Performance: 12.5874 MHz sprite line buffer render clock, 6.2937 MHz sprite line buffer scan/erase & pixel clock[54]

- Display resolution: 320×224 to 400×262 pixels,[74] progressive scan (non-interlaced)

- Frame rate: 60 frames per second[78]

- Color palette: 32,768 (*Hang-On*), or 98,304 (*Space Harrier, Enduro Racer*)[57]

- Colors on screen: 6144[74]

- Graphical planes:[75]

  - 2 tilemap layers: Row & column scrolling[58]

  - Text layer

  - Sprite layer: Hardware sprite-scaling

  - Road layer: 512×256 resolution[72]

  - Translucent shadows (*Space Harrier*)[20]

- Sprite capabilities: Hardware sprite-scaling, 128 sprites on screen per frame,[78] thousands of sprites scaled per second,[79] dual line buffers, double buffering,[54] 800 sprite pixels (800.75 sprite processing ticks) per scanline, 100 sprites per scanline,[54] 8[56] to 256[54] width, 8[56] to 256[54] height

## Sega OutRun

*Sega OutRun* was a 16-bit arcade system released in 1986 for the driving game *Out Run* (1986). It was also used for *Super Hang-On* (1987) and *Turbo Outrun* (1989). It is the second in Sega's Super Scaler series of pseudo-3D arcade hardware.

### Specifications

- Main CPU: 2× Motorola 68000 @ 12.5 MHz[80] (16-bit & 32-bit instructions @ 4.375 MIPS)[6]

- Sound CPU: Zilog Z80 @ 4 MHz[80] (8-bit & 16-bit instructions @ 0.58 MIPS)[6]

- Sound chips:

  - FM synthesis chip: Yamaha YM2151 @ 4 MHz (8 FM channels)

  - PCM chip: SegaPCM (315-5218[81]) @ 4 MHz[76] (stereo output, 16 PCM channels, 12-bit audio,[77] 31.25 kHz sampling rate[75])

- GPU: Sega Super Scaler chipset

  - Graphics board: Sega 837-6064 / 171-5377 VIDEO Board[20] @ 25.1748 MHz[76] (315-5197 Sega Custom Tilemap Generator, 315-5211 Sega Custom Sprite Generator, 315-5242 Sega Custom Color Encoder)[76]

  - Road graphics chips: 315-5155 Sega Road Bit Extraction, 315-5222 Signetics PLS153N Road Mixing[76]

- Display resolution: 320×224 to 400×262,[76] progressive scan

- Refresh rate: 60.0543 Hz (V-sync)[76]

- Frame rate: 30 frames per second[82]

- Color palette: 98,304[57]

- Colors on screen: 12,288[76]

- Graphical planes:[80]

  - 2 tilemap layers: System 16B tilemap system, row & column scrolling,[20] parallax scrolling[83]

  - 1 text layer

  - 1 sprite layer: Hardware sprite-scaling/zooming

  - 1 road layer: Can draw 2 roads at once, 512×256 pixels each,[72] tiled bitmaps[84]

  - Translucent shadows

- Sprite capabilities: Framebuffered sprites with zooming capabilities,[20] 128 on-screen sprites per frame,[80] thousands of sprites scaled per second,[79] 16 colors per sprite[85]

- Sprite pixels: 25.1748 MHz video clock cycles (60.0543 Hz refresh rate),[76] 419,199 pixels per screen refresh (262 scanlines), 1600 sprite pixels per scanline, 128 sprites per scanline

## Sega X Board

For the military planning calendar, see X-board.

The *Sega X Board* is an arcade system board released by Sega in 1987. As the third in Sega's Super Scaler series of arcade hardware, it was noteworthy for its sprite manipulation capabilities, which allowed it to create high quality pseudo-3D visuals. This trend would continue with the Y Board and the System 32, before the Model 1 made true 3D arcade games more financially affordable.

## X Board specifications

- Main CPU: Hitachi FD1094 (Motorola 68000) @ 12.5 MHz,[86] Motorola MC68000 @ 12.5 MHz[87] (16-bit & 32-bit instructions @ 4.375 MIPS)[6]

- Sound CPU: Zilog Z80 @ 4 MHz[87] (8-bit & 16-bit instructions @ 0.58 MIPS)[6]

- Sound chips:[87]

    - FM synthesis chip: Yamaha YM2151 @ 4 MHz (8 FM channels)

    - PCM chip: SegaPCM (315-5218) @ 4 MHz[86] (stereo output, 16 PCM channels, 12-bit audio,[77] 31.25 kHz sampling rate[75])

- GPU: Sega Super Scaler chipset @ 50 MHz[86]

    - Main graphics chips: 315-5197 tilemap generator, 315-5211A sprite generator, 315-5242 color encoder, 315-5275 road generator, 315-5278 sprite ROM bank control[20]

    - Math chips:[58] 315-5248 hardware multiplier, 315-5249 hardware divider[86]

- Display resolution: 320×224[87] to 400×262,[77][86] progressive scan

- Refresh rate: 59.6368[87] to 60[88] Hz (V-sync)

- Frame rate: 59.6368[87] to 60[88] frames per second

- Board composition: Single board

- Color palette: 98,304[57]

- Colors on screen: 24,576[86]

- Graphical planes:[87]

    - 4 tile layers

    - 1 text layer

    - 1 sprite layer with hardware sprite zooming

    - 1 road layer, can draw 2 roads at once

    - Translucent shadows

- Sprite capabilities: Dual sprite framebuffers, 512×256 framebuffer resolution,[77] hardware sprite zooming,[87] sprite rotation,[88] thousands of sprites scaled per second[79]

    - Sprite size: 8×8[86] to 512×256[77] pixels

    - Colors per sprite: 16[77]

    - Sprites per frame: 256 on screen at one time[87]

    - Sprite pixels: 50 MHz video clock cycles,[86] 833,333 (60 Hz) to 838,408 (59.6368 Hz) pixels per frame (262 scanlines), 3180 to 3200 sprite pixels per scanline, 256 sprites per scanline

*Super Monaco GP* (1989) added the following upgrades:[86]

- Additional boards: Network Board, Sound Board, Motor Board

- Additional CPU: 2× Zilog Z80 @ 8 MHZ (2.32 MIPS)

- Additional sound CPU: Zilog Z80 @ 4 MHz (0.58 MIPS)

- Additional sound chip: SegaPCM @ 4 MHz[86] (additional 16 PCM channels,[77] totalling 32 PCM channels)

- Sound output: 4-channel surround sound[86]

## Sega Y Board

The *Sega Y Board* is an arcade system board released by Sega in 1988. Like the X Board before it, the Y Board was known for its pseudo-3D sprite manipulation capabilities, handled by Sega's custom Super Scaler chipset.

**Y Board specifications**

- Board composition: CPU Board + Video Board

- Main CPU: 3× MC68000 @ 12.5 MHz[89] (16-bit & 32-bit instructions @ 6.563 MIPS)[6]

- Sound CPU: Z80 @ 4 MHz[89] (8-bit & 16-bit instructions @ 0.58 MIPS)[6]

- Sound chips: YM2151 @ 4 MHz, SegaPCM @ 15.625 kHz

- Sound chips:[89]

  - FM synthesis chip: Yamaha YM2151 @ 4 MHz (8 FM channels)

  - PCM chip: SegaPCM (315-5218) @ 4 MHz (stereo output, 16 PCM channels, 12-bit audio,[77] 31.25 kHz sampling rate[75])

- GPU: Sega Super Scaler chipset[90]

  - Graphics board: Sega 837-6566 Video Board[20] @ 50 MHz[90] (315-5196 sprite generator, 315-5213 sprite chip, 315-5242 color encoder, 315-5305 sprite generator, 2× 315-5306 video sync & rotation, 315-5312 video mixer)[20]

  - Math chips:[58] 315-5248 hardware multiplier, 315-5249 hardware divider[20]

- RAM: 778 KB (SRAM[86])

  - Main RAM: 208 KB (64 KB CPU 1, 16 KB CPU 2, 64 KB CPU 3, 64 KB shared)[90]

  - Video RAM: 566 KB (32 KB Y-sprites, 4 KB B-sprites, 2 KB rotation, 16 KB palette,[90] 512 KB framebuffer[77])

  - Sound RAM: 6 KB (2 KB Z80,[90] 4 KB SegaPCM[86])

- Display resolution: 320×224[89] to 342×262,[90] progressive scan

- Sprite resolution: Up to 512×512 pixels[91]

- Refresh rate: 59.6368[87] to 60[90] Hz (V-sync)

- Frame rate: 59.6368[87] to 60[90] frames per second

- Color palette: 2,097,152 (4096 palette banks with 512 colors each),[89] to 16,777,216 with effects (shadow & highlight, luminosity, palette fade)

- Colors on screen: 24,576,[90] to 98,304 with luminosity and palette fade

- Graphical planes: Three layers[20][89]

  - B-sprite (front plane) layer: Priority on top, based on System 16B (line buffer[58]) sprite system

  - Y-sprite (back plane) layer: Plugs into a full-screen rotation, large fillrate, dual framebuffers[20] (based on X Board[77]) that can be fully rotated

  - Sky gradient (background) layer: Bitmap plane

- Sprite capabilities: Linked list of sprites,[89] shadow & highlight,[58] palette fade,[90] color rotations, different levels of luminosity, full sprite zooming & scaling on both sprite planes,[89] full sprite & frame-buffer rotation on Y-sprite plane,[20] double buffering, dual line buffers on B-plane (512 sprite pixels per line),[58] dual framebuffers on Y-plane[20]

  - Sprite size: 8×8[90] to 512×512[91] pixels

  - Colors per sprite: 16 to 512[89]

  - Sprites per frame: 68 KB sprite RAM,[90] up to 2176 sprites (with 8x8 size and 16 colors each)

  - Sprite pixels: 50 MHz video clock cycles,[90] 833,333 (60 Hz) to 838,408 (59.6368 Hz) pixels per frame (262 scanlines), 3180 to 3200 sprite pixels per scanline, 397 to 400 sprites per scanline

## 4.2.11 Sega Mega series

**Sega Mega-Tech**

The *Sega Mega-Tech* was an arcade system developed by Sega Europe in 1988. It is based on the Mega Drive/Genesis video game console hardware, and more or less identical.[92] Its operation ability is similar to Nintendo's PlayChoice-10, where the credits bought give the user a playable time period rather than lives (usually 1 minute per credit), and can switch between games during playtime.

A few things were omitted, such as the expansion hardware allowing for Sega Mega-CD or Sega 32X as these were not developed at this point, so would not likely be offered as an arcade expansion. The PCB for the Mega-Tech also includes the ability to display to a second monitor, which contains a list of the games installed in the machine and also displays instructions for controlling the game, 1 or 2 player information, and a short synopsis of each game. The second monitor also displays the time left for playing.

Since the machine was basically a Mega Drive with timer control for arcade operations, porting games to the Mega-Tech was an easy task and so many games were released, most of them popular titles such as Streets Of Rage, Revenge Of Shinobi, Golden Axe, Sonic The Hedgehog and many more. The ability was also added for the machine to play Sega Master System titles, though fewer Master System titles were ported than Mega Drive titles. Some include the original Shinobi, Outrun and After Burner.*[93]

The Sega Mega-Tech was released in Europe, Australia, and Asia (including Japan), but not in North America.

**Sega Mega-Play**

The Sega Mega-Tech system was soon replaced by its successor, the Mega-Play, a JAMMA based system.*[92] This system utilized only 4 carts instead of 8. This version also utilizes traditional arcade operations, in which credits bought are used to buy lives instead.*[94]

Like the Mega-Tech, The Sega Mega-Play was released in Europe, Australia, and Asia (including Japan), but not in North America.

**Sega System 14 / C / C-2**

Sega's *System 14*, also known as *System C* and *System C-2*, is a Jamma PCB used in arcade games, introduced in 1989. This hardware is based closely on the Sega Mega Drive/Genesis hardware, with the main CPU, sound processor and graphics processor being the same,*[95] but with the addition of the Altera EPM5032*[96] and Sega 315-5242 color encoder*[20] increasing the color palette. The CPU clock speed is slightly faster (8.94 MHz instead of 7.67 MHz), there is no Z80, and the sound chip is driven by the CPU. The DAC is also replaced by the NEC μPD7759, the same as the System 16 hardware. 17 known games were created for the System C-2 hardware.

**Specifications**

- Board composition: Single JAMMA board*[95]

- Main CPU: MC68000 @ 8.948862 MHz*[95] (16-bit & 32-bit instructions @ 1.566 MIPS)*[6]

- Sound chip: YM3438 @ 7.670453, SN76496 @ 3.579545

- Optional sound chip: NEC μPD7759 @ 640 kHz*[95] (9-bit ADPCM @ 8 kHz sampling rate)*[97]

- Graphics chips: Yamaha YM7101 VDP, Altera EPM5032,*[96] Sega 315-5242 color encoder*[20]

- Video resolution: 320×224 pixels

- Color palette: 98,304*[57]

- Colors on screen: 6144*[96]

- Hardware features: Line scroll, column scroll, raster interrupt, 2 background planes (one with an option window), sprite plane, several levels of priority

### 4.2.12   Sega System 32

*System 32* was an arcade platform released by Sega in 1990. It succeeded the Y Board and System 24, combining features from both. It used a NEC V60 processor at 16.10795 MHz, supporting 32-bit fixed-point instructions as well as 32-bit and 64-bit floating-point instructions. It used a new custom Sega graphics chipset combining the Y Board's pseudo-3D Super Scaler capabilities with the System 24's sprite rendering system. Notable titles included *Golden Axe: The Revenge of Death Adder*, *Rad Mobile*, *OutRunners*, and *SegaSonic the Hedgehog*.

There was another version of the System 32 hardware, called *System Multi 32* or *System 32 Multi*, released in 1992. This was similar to the original, but had a dual monitor display, a new NEC V70 processor at 20 MHz, a new Sega MultiPCM sound chip, more RAM, and other improvements. This was the last of Sega's Super Scaler series of pseudo-3D arcade system boards.

**System 32 specifications**

- Main CPU: NEC V60 @ 16.10795 MHz*[98]

  - Fixed-point arithmetic: 32-bit RISC*[99] instructions @ 3.524 MIPS (million instructions per second)*[100]

  - Floating-point unit: 32-bit and 64-bit operations*[101]

- Sound CPU: Zilog Z80 @ 8.053975 MHz*[98] (8-bit & 16-bit instructions @ 1.168 MIPS*[6])

- Sound chips:

  - FM synthesis chips: 2× Yamaha YM3438 (based on Yamaha YM2612) @ 8.053975 MHz (12 FM channels)

  - PCM sampling chip: Ricoh RF5c68 @ 12.5 MHz (8 PCM channels)

- GPU: Sega Super Scaler 317-5964 chipset (315-5242 video DAC/color encoder, 315-5386 tilemap generator, 315-5387 sprite generator, 315-5388 video mixer/color blender)*[20]*[98]

- RAM: 1684.125 KB

- V60 main RAM: 584 KB (64 KB work, 8 KB shared, 512 KB random number generator)*[102]

- V60 video RAM: 320.125 KB (128 KB video, 128 KB sprite attributes, 64 KB palette, 128 bytes mixer)*[102]

- Framebuffer DP VRAM: 768 KB (16× 32 KB Hitachi HM53461ZP-12,*[102]*[103] 8× 32 KB NEC uPD42264*[102]*[104])

- Z80 sound RAM: 12 KB (4 KB RF5c68, 8 KB shared)*[102]

- Display resolution: 320×224 to 416×262 pixels,*[102] progressive scan

- Frame rate: 60 frames per second*[102]

- Graphical capabilities: Color rotations, different levels of luminosity,*[98] 7 levels*[105]*[106] of global RGB brightness control,*[98]*[107] fading & lighting,*[108] shadow & highlight, 8 levels of alpha blending, tile flipping, line & row scrolling,*[106] palette indirection, dynamic priorities, per-color priority, per-component color control*[20]

- Color palette: 2,097,152 (4096 palette banks with 512 colors each*[98]) to 16,777,216 (with shadow & highlight and 7 levels of RGB brightness control)

- Colors on screen: 49,152 (16,384*[98] with shadow & highlight*[106]) to 786,432 (with luminosity and 8 levels of alpha blending)

- Graphical planes:

  - 4 tilemap*[106] background planes: Scaling, line-scrolling,*[98] line selection, line zoom, alpha blending, window clipping*[20]

  - 1 tilemap text layer*[20]

  - 1 bitmap layer*[106]

  - 1 background layer*[106]

  - 2 sprite layers*[106]

- Sprite capabilities: Linked lists of sprites,*[20] double buffering, dual framebuffers,*[106] technically infinite sprites of arbitrary size, sprite-scaling,*[98] sprite rotation,*[109]*[110] jumping & clipping capabilities, advanced hot-spot positioning,*[20] System 24 sprite rendering system*[65]

  - Sprite size: 8*[65] to 2048 pixels in width/height*[106]

  - Colors per sprite: 16 to 512*[98]

- Sprites per frame: 128 KB sprite attribute RAM,*[102] 16 bytes per sprite,*[106] 8192 sprites per frame

- Sprite pixels per scanline: 4096*[65]*[106]

- Sprites per scanline: 512

**System Multi 32 specifications**

Sega System Multi 32 included the following upgrades in 1992:

- Main CPU: NEC V70 @ 20 MHz*[111]

  - Fixed-point arithmetic: 32-bit RISC instructions @ 6.6 MIPS*[100]

  - Floating-point unit: 32-bit and 64-bit operations*[101]

- Sound CPU: 2× Zilog Z80 @ 8.053975 MHz (8-bit & 16-bit instructions @ 2.336 MIPS*[6])

- Sound chips:

  - FM synthesis chip: Yamaha YM3438 @ 8.053975 MHz (6 FM channels)

  - PCM sampling chip: Sega MultiPCM*[105] (28 PCM channels)

- GPU: 2× Sega Super Scaler 317-5964 chipset

- Display resolution: Dual monitor,*[111] 640×448 to 832×262 pixels, progressive scan

- Color palette: 4,194,304 (2,097,152 per screen) to 16,777,216 (with shadow & highlight and RGB brightness control)

- Colors on screen: 98,304 (49,152 per screen) to 1,572,864 (786,432 per screen)

- Graphical planes: 4 sprite layers*[106]

- Sprite capabilities: Multiple buffering, 4 framebuffers*[106]

### 4.2.13 Sega Model series

**Sega Model 1**

The *Sega Model 1* is an arcade system board released by Sega in 1992. It was Sega's first polygonal 3D hardware. The first game for the system, *Virtua Racing*, was designed to test the viability of the platform and was never intended to be released commercially, but it was such a success internally that Sega did so anyway.

However, the high cost of the Model 1 system meant only six games were ever developed for it, among them the popular fighting game *Virtua Fighter*. Like the previous Super Scaler pseudo-3D arcade boards, the Model 1 3D arcade board was designed by Sega AM2's Yu Suzuki.*[73]

## Model 1 specifications

- Main CPU: NEC V60 @ 16 MHz

  - Fixed-point arithmetic: 32-bit RISC*[99] instructions @ 3.5 MIPS (million instructions per second)*[100]*[112]

  - Floating-point unit: 32-bit and 64-bit operations*[101]

- Graphics board: Sega 837-7894 171-6080D VIDEO PCB*[113]

- GPU coprocessors: 5× Fujitsu TGP MB86233*[114] (geometrizer, rasterizer,*[115] DSP, FPU)

  - Coprocessor abilities: Floating decimal point operation function, axis rotation operation function, 3D matrix operation function*[99]

  - Floating-point unit: 32-bit operations @ 16 MFLOPS (Mega-FLOPS)*[99]*[116] each (80 MFLOPS combined)

- Sound CPU: Motorola 68000 @ 12 MHz*[99]

- Sound chips: 2× Sega 315-5560 Custom Multi-PCM*[99]

  - Audio capabilities: 28 PCM channels per chip (one for music, one for sound effects), 56 PCM channels total

- Sound timer: Yamaha YM3834 @ 8 MHz

- RAM: 1936 KB (1880 KB SRAM)*[113]

  - Main SRAM: 408 KB

  - Video memory: 1464 KB SRAM (192 KB display list, 576 KB tiles, 64 KB colors)

  - Audio memory: 64 KB (8 KB SRAM)

- Monitor display resolution: 496 × 384 pixels, 24 kHz horizontal sync,*[113] 60 Hz refresh rate, progressive scan (non-interlaced)

- Frame rate: 60 frames per second*[117]

- Colors: 16,777,216 (16-bit high color depth*[99] and 256 luminance levels)*[115]

- Graphical capabilities: Shading, flat shading, diffuse reflection, specular reflection, 2 layers of background scrolling, alpha blending, alpha channel,*[99] lighting*[115]

- Geometric performance: 180,000 polygons/sec (with all effects), 540,000 vectors/sec*[99]

- Rendering fillrate: 1,200,000 pixels/sec*[99]

## Sega Model 2

The *Sega Model 2* is an arcade system board released by Sega in 1993. Like the Model 1, it was developed in cooperation with Martin Marietta, and was a further advancement of the earlier Model 1 system. The most noticeable improvement was texture mapping, which enabled polygons to be painted with bitmap images, as opposed to the limited monotone flat shading that Model 1 supported. The Model 2 also introduced the use of texture filtering and texture anti-aliasing.*[118]

Designed by Sega AM2's Yu Suzuki, he stated that the Model 2's texture mapping chip originated "from military equipment from Lockheed Martin, which was formerly General Electric Aerial & Space's textural mapping technology. It cost $2 million to use the chip. It was part of flight-simulation equipment that cost $32 million. I asked how much it would cost to buy just the chip and they came back with $2 million. And I had to take that chip and convert it for video game use, and make the technology available for the consumer at 5,000 yen ($50)" ($84 in 2015) per machine. He said "it was tough but we were able to make it for 5,000 yen. Nobody at Sega believed me when I said I wanted to purchase this technology for our games." There were also issues working on the new CPU,*[73] the Intel i960-KB, which had just released in 1993.*[119] Suzuki stated that when working "on a brand new CPU, the debugger doesn't exist yet. The latest hardware doesn't work because it's full of bugs. And even if a debugger exists, the debugger itself is full of bugs. So, I had to debug the debugger. And of course with new hardware there's no library or system, so I had to create all of that, as well. It was a brutal cycle." *[73]

Despite its high pricetag, the Model 2 platform was very successful. It featured some of the highest grossing arcade games of all time: *Daytona USA,*[120] *Virtua Fighter 2*, *Cyber Troopers Virtual-On*, *The House of the Dead*, and *Dead or Alive*, to name a few.

Model 2 has four different varieties: Model 2 (1993),*[120] Model 2A-CRX*[121] (1994),*[122] Model 2B-CRX*[123] (1994)*[124] and Model 2C-CRX (1996).*[125] While Model 2 and 2A-CRX use a custom DSP with internal code for the geometrizer, 2B-CRX and 2C-CRX use well documented DSPs and upload the geometrizer code at startup to the DSP. This, combined with the fact that some games were available for both 2A-CRX and 2B-CRX, led to the reverse engineering of the Model 2 and Model 2A-CRX DSPs.

**Model 2 specifications**

**Main CPU (central processing unit)**

- Main CPU: Intel i960-KB @ 25 MHz

    - Fixed-point arithmetic: 32-bit RISC instructions @ 25 MIPS (million instructions per second)[126]

    - Floating-point unit: 32-bit, 64-bit and 80-bit operations @ 13.6 MFLOPS (Mega-FLOPS, or million floating-point operations per second) (Whetstone)[119]

**GPU (graphics processing unit) video hardware**

- Geometry Engine[127] DSP coprocessors: 6× Fujitsu TGP MB86234 (Model 2/2A-CRX),[114] or 2× Analog Devices ADSP-21062 SHARC (Model 2B-CRX), or 2× Fujitsu TGPx4 MB86235 (Model 2C-CRX)[114]

    - Coprocessor abilities: Floating decimal point operation function, axis rotation operation function, 3D matrix operation function

    - Floating-point unit:

        - Model 2/2A-CRX: 32-bit operations @ 16 MFLOPS[121] ×6 (96 MFLOPS)

        - Model 2B/2C-CRX: 32-bit & 40-bit operations @ 120 MFLOPS[128] ×2 (240 MFLOPS)

    - Fixed-point arithmetic: 32-bit & 48-bit instructions @ 80 MIPS (Model 2B-CRX)[128]

- Hardware Renderer:[127] Sega-Lockheed-Martin Custom rasterization[129] & texture mapping hardware (Model 2),[73] or 2× Fujitsu MB86271 AGP (Model 2C-CRX)[130]

    - Fixed-point arithmetic: 32-bit & 64-bit instructions @ 240 MIPS (Model 2C-CRX)[131]

- Z-Sort & Clip Hardware[127] (2× Fujitsu MB86272 Z-sorter in Model 2C-CRX)[130]

- Sega System 24 tilemap engine[129]

**Audio hardware**

- Sound CPU: Motorola 68000 @ 10 MHz (Model 2), or Motorola 68000 @ 12 MHz (Model 2A/2B/2C-CRX)

- Sound chip: 2× Sega 315-5560 Custom MultiPCM (Model 2), or Yamaha SCSP (Model 2A/2B/2C-CRX)

- Sound timer: Yamaha YM3834 @ 8 MHz (Model 2 only)

- PCM channels: 56[120]

- PCM sample ROM: 16 Mbits (Model 2),[120] or 68 Mbits (Model 2A/2B/2C-CRX)[121]

- PCM quality: 16-bit depth,[132] 44.1 kHz sampling rate (CD quality)[121]

- SCSP features: 128-step DSP, 32 FM synthesis channels, 32 MIDI channels, 32 LFO channels[132]

**RAM (random access memory)**

Total RAM: 9776 KB (Model 2/2A-CRX), or 18,388 KB (Model 2B/2C-CRX)

- Main RAM: 1152 KB (9 Mbits)[123] (1024 KB work, 64 KB network, 64 KB serial)[129]

- Video memory: 5984 KB (Model 2/2A-CRX), or 14,596 KB (Model 2B/2C-CRX)

    - Framebuffer VRAM:[131] 1024 KB (Model 2/2A-CRX), or 1536 KB (Model 2B/2C-CRX)[129]

    - Coprocessor buffer[129] SRAM/SDRAM:[131] 64 KB (Model 2/2A-CRX), or 8228 KB (Model 2B/2C-CRX)[129]

    - Texture memory: 4096 KB[129] SRAM/SDRAM[131]

    - Luma: 128 KB (Model 2/2A-CRX), or 64 KB (Model 2B/2C-CRX)[129]

    - Other: 672 KB (32 KB geometry, 576 KB tiles, 64 KB colors)[129]

- Audio memory: 576 KB[129]

- Backup SRAM/NVRAM: 16 KB[129]

- Extra RAM: 2048 KB[129]

**Graphical capabilities**

- Monitor display resolution: 496 × 384 pixels, 24 Hz horizontal sync, 60 Hz refresh rate, progressive scan (non-interlaced)[123]

- Texture map resolution: Up to 1024 × 2048 pixels[123]

    - Microtexture size: Up to 128 × 128 pixels

- Color depth: 16,777,216 (24-bit true color)[123][133][134]

- Graphical features: Flat shading, texture mapping, perspective correction, texture filtering, texture anti-aliasing, microtexture, diffuse reflection, specular reflection, alpha blending, transparency,[123] rasterization, mipmapping, level of detail,[127] z-sorting and T&L (transform, clipping, and lighting)[127][130]

    - Model 2C-CRX: Gouraud shading, hidden surface, z-buffering,[130] point sampling, bilinear filtering, trilinear filtering[134]

- Frame rate: 60 frames/sec[120]

- Geometric performance:

    - Model 2: 300,000 textured quad polygons/sec[120] to over 500,000 textured triangle polygons/sec,[133] 900,000 vectors/sec[120]

    - Model 2C-CRX: 490,000 textured polygons/sec (with clipping, lighting and Gouraud shading)[135] to 900,000 textured triangle polygons/sec (with Gouraud shading)[131]

- Rendering fillrate:

    - Pixel fillrate: 1.2 million pixels/sec (Model 2)[120] to 120 million pixels/sec (2 million pixels/frame) (Model 2B-CRX/2C-CRX)[123][125]

    - Texture fillrate: 36 million texels/sec (500 pixels/polygon) (Model 2C-CRX)[136]

## Sega Model 3

The *Sega Model 3* is an arcade system board released by Sega in 1996. It was the final culmination of Sega's partnership with Lockheed Martin, using the company's Real3D division to design the graphical hardware. Upon release, the Model 3 was easily the most powerful arcade system board in existence,[137] capable of over one million quad polygons per second and over two million triangular polygons per second.[138] The hardware went through several "steppings," which increased the clock speed of the CPU and the speed of the 3D engine, as well as minor changes to the board architecture.[139] Step 1.0 and Step 1.5 released in 1996,[138][140] Step 2.0 in 1997,[141] and Step 2.1 in 1998.[142]

Well known Model 3 games include *Virtua Fighter 3* (1996), *Sega Super GT* (1996), *Harley-Davidson & L.A. Riders* (1997), *Sega Bass Fishing* (1997), *Daytona USA 2* (1998), *Sega Rally 2* (1998), and *The Ocean Hunter* (1998), although it is the rarest of them. By 2000, the Sega Model 2 & 3 had sold over 200,000 arcade systems worldwide,.[143]

## Model 3 specifications

- Main CPU: IBM-Motorola PowerPC 603e[137] (32-bit & 64-bit instructions)[144]

    - Step 1.0: 66 MHz[138] (93.4 MIPS,[144] 132 MFLOPS)[145]

    - Step 1.5: 100 MHz[140] (142 MIPS,[144] 200 MFLOPS)[145]

    - Step 2.0: 166 MHz[141] (235 MIPS,[144] 332 MFLOPS)[145]

- Sound CPU : Motorola 68000 @ 12 MHz[139] (2.1 MIPS)[6]

- Sound chips: 2× Yamaha SCSP/YMF292-F[137]

    - PCM audio: 64 voices/channels, 16-bit depth, 44.1 kHz sampling rate (CD quality)[137][138]

    - Other features: 128-step DSP, 32 FM synthesis channels, 32 MIDI channels, 32 LFO channels,[132] 4-channel surround sound,[137] 16.5 MB audio ROM[138]

- Optional sound board: MPEG Sound Board[138]

    - Sound CPU: Motorola 68000 or Zilog Z80

    - Sound chip: NEC uD65654GF102

    - Features: MPEG audio compression, stereo output, steam individual mono channels to left and right speakers

- Video board:[139]

    - Step 1.0: Sega 837-11859 MODEL3

    - Step 1.5: Sega 837-12875 MODEL3 STEP 1.5

    - Step 2.0: Sega 837-12716 MODEL3 STEP2

    - Step 2.1: Sega 837-13368 MODEL3 STEP2.1

- GPU: 2× Lockheed Martin Real3D/Pro-1000

    - Texture mapping: Mipmapping, perspective correction,[137] texture filtering[138]

- Anti-aliasing:[137] Texture anti-aliasing, multi-layered anti-aliasing[138]

- Shading: Flat shading, Gouraud shading, high-specular Gouraud shading, micro texture shading,[137] fix shading[138]

- Lighting: Parallel light, pin-point light, 4 light spots,[137] 4 spot lights[138]

- Other special effects: Zoning fog, 32 levels of translucency, clipping, model & texture LOD, fade in/out, 4095 moving models,[137]

- Other capabilities: T&L (transform, clipping, and lighting),[137][146] alpha blending, trilinear filtering, trilinear interpolation, specular reflection, specular highlight,[137] z-buffering,[137] culling[147]

- ALU: Mitsubishi 3D-RAM[139][148]

  - Framebuffer resolution: 1280×1024[139]

  - Capabilities: Blending, depth check, stencil & raster operations,[148] pixel buffer, tiled rendering,[139] z-compare, alpha blending, up to 400 million pixels/sec rendering fillrate[149]

- Monitor display resolution: 496×384[138] to 640x480,[137] progressive scan (non-interlaced)[137]

  - Refresh rate: 60 Hz,[139] 60 frames per second

- Color depth: ARGB,[147] 24-bit RGB[150] true color (16,777,216 colors) and alpha opacity

- Geometric performance: 1,000,100 textured quad polygons/sec, 2,000,200 textured triangle polygons/sec,[138] with all effects (Step 1.0)

- Rendering fillrate: 60 million[137] to 400 million[149] pixels/sec, 16 million coloured textures/sec[137]

RAM: 33,321 KB

- Main RAM: 8192 KB[139] (8 MB)[138]

- Video RAM: 23,713 KB (8 MB texture memory, 1 MB display list, 4 MB culling, 4 MB polygons,[147] 5 MB framebuffer 3D-RAM, 1152 KB tilemap generator VRAM, 33 KB SRAM cache)[139]

  - 4× Mitsubishi 3D-RAM: 5 MB (4× 1.25 MB) fast framebuffer SD VRAM, 1 KB (4× 256 bytes) pixel buffer SRAM cache[139][149][151]

- 8× Hitachi HM5241605 SDRAM: 4 MB (8× 512 KB)[139][152]

- 16× Mitsubishi M5M4V4169 cache: 8 MB (16× 512 KB) SDRAM, 32 KB (16× 2 KB) SRAM[139][153]

- Audio RAM: 1096 KB (64 KB main, 1032 KB SCSP)[139]

- Other RAM: 320 KB (192 KB security, 128 KB backup static NVRAM)[139]

## 4.2.14  Sega ST-V

*Sega ST-V PCB*

*ST-V* (*Sega Titan Video game system*) was an arcade system board released by Sega in 1994.[154] Departing from their usual process of building custom arcade hardware, Sega's ST-V is essentially identical to the Sega Saturn home console system. The only difference is the media: ST-V used ROM cartridges instead of CD-ROMs to store games. Being derived from the Saturn hardware, the ST-V was presumably named after the moon Titan, a satellite of Saturn.

The majority of ST-V titles were released in Japan only, but a notable exception was the port of *Dynamite Deka*, which became *Die Hard Arcade*. Games released for the ST-V includes the arcade version of *Virtua Fighter Remix*, *Golden Axe: The Duel* and *Final Fight Revenge*. The shared hardware between Saturn and ST-V allowed for very "pure" ports for the Saturn console.

**ST-V specifications**

- Main CPU processors: 2× Hitachi SH-2 (7604 32-Bit RISC) @ 28.6 MHz, in a master/slave configuration

  - Fixed-point arithmetic: 32-bit RISC instructions @ 28 MIPS each, 56 MIPS combined[155]

- DSP coprocessor: Custom Saturn Control Unit (SCU)[156]

  - Fixed-point arithmetic: Up to 4 parallel instructions

- VDP1: 32-bit Video Display Processor, handles sprite/texture and polygon drawing[156]

  - Framebuffers: Dual 256 KB framebuffers with rotation & scaling,[156] three framebuffer sizes (512×256, 512×512, 1024×256)[157]

  - 3D polygon capabilities: Texture mapping, shading, flat shading, Gouraud shading[156]

    - Polygon rendering performance: 200,000 texture-mapped polygons per second, 500,000 flat-shaded polygons per second[156]

  - Sprite/Texture capabilities: Rotation & scaling,[156] flipping, distortion,[157] virtually unlimited color tables, virtually unlimited sprites,[158] System 24 sprite rendering system[65]

    - Sprite/Texture memory cache: 512 KB[156][159]

    - Sprite/Texture size: 8×1 to 512×255 pixels[160]

    - Colors per sprite/texture: 16, 64, 128, 256, and 32,768[161]

    - Sprites/Textures per frame: 512 KB sprite/texture memory, 32 bytes per sprite/texture,[162] 16,384 sprites/textures per frame

    - Sprite/Texture pixels/texels per line: 4096[65]

    - Sprites/textures per line: 512

  - Other features: Alpha blending, clipping, luminance, shadows, transparency,[157] anti-aliasing[163]

- VDP2: 32-bit Video Display Processor, handles background and scroll planes[156]

  - Features: Transparency effects, shadowing, 2 windows for special calculations, 5 simultaneous scrolling backgrounds, 2 simultaneous rotating playfields, background scaling[156]

- Tilemap planes: Up to 4 scrolling tilemaps @ 512×512 to 1024×1024 pixels and 2 rotating tilemaps @ 512×256 to 1024×512 pixels, two tile sizes (8×8 and 16×16), column/row/line scrolling[164]

- Bitmap planes: Up to 2 scrolling bitmaps @ 512×256 to 1024×512 pixels and 1 rotating bitmap @ 512×256 to 512×512 pixels[164]

- Sound CPU: Motorola 68000 @ 11.45456 MHz[156]

- Sound chip: Yamaha YMF292-F SCSP @ 11.3 MHz[132][155][156]

  - PCM audio: 32 channels, 16-bit depth, 44.1 kHz sampling rate (CD quality)

  - Other features: 128-step DSP, 32 FM synthesis channels, 32 MIDI channels, 32 LFO channels

- Main RAM: 4.04 MB[156]

  - Main RAM: 2 MB

  - VRAM: 1.54 MB (including dual 256 KB framebuffers, 512 KB texture cache, and 512 KB background VRAM)[155]

  - Audio RAM: 512 KB[156]

- Display resolution: 320×224 to 720×576[156]

- Frame rate: Up to 60 frames per second[156]

- Colors: 16,777,216 (24-bit true color) on screen,[156] up to 32,768 (15-bit high color) per sprite/texture,[157] up to 16,777,216 colors per background[164]

### 4.2.15  Sega NAOMI series

**Sega NAOMI**

First demonstrated in November 1998 at JAMMA, since just before the release of The House of the Dead 2 in Japan. The Sega *Naomi* (*New Arcade Operation Machine Idea*) is the successor to the Sega Model 3 hardware.

A development of the Dreamcast home game console, the NAOMI and Dreamcast share the same hardware components: Hitachi SH-4 CPU, PowerVR Series 2 GPU (PVR2DC), and Yamaha AICA Super Intelligent Sound Processor based sound system. NAOMI has twice as much system memory, twice as much video memory, and four times as much sound memory.

Multiple NAOMI boards can be 'stacked' together to improve graphics performance, or to support multiple-monitor output. A special game cabinet for the NAOMI,

NAOMI Universal Cabinet, houses up to sixteen boards for this purpose. Multiple-board variants are referred to as *NAOMI Multiboard* hardware, which debuted in 1999.[165] Games of this type became a standard with the introduction of large-scale satellite arcade machines with physical card elements that link up multiple boards such as *Derby Owners Club* and *World Club Champion Football*.

The other key difference between NAOMI and Dreamcast lies in the game media. The Dreamcast reads game data from GD-ROM optical disc, while the NAOMI arcade board features 168 MB of solid-state ROMs or GD-ROMs using a custom DIMM board and GD-ROM drive. In operation, the NAOMI GD-ROM is read only once at system power up, loading the disc's contents to the DIMM Board RAM. Once loading is complete, the game executes only from RAM, thereby reducing mechanical wear on the GD-ROM drive.

Unlike Sega's previous arcade platforms (and most other arcade platforms in the industry), NAOMI is widely licensed for use by other game publishers including Sega, Namco Bandai, Capcom, Sammy and Tecmo Koei. Games such as *Mazan*, *Marvel Vs. Capcom 2*, *Dead or Alive 2* and *Guilty Gear XX* were all developed by third-party licensees of the NAOMI platform. An offshoot version of the NAOMI hardware is Atomiswave by Sammy Corporation.

After nine years of hardware production, and with new game titles coming in 2008 like Melty Blood: Actress Again and Akatsuki Blitzkampf AC, NAOMI is considered to be one of the longest running arcade platforms ever and is comparable in longevity with the Neo-Geo MVS.

**NAOMI specifications**

- CPU: Hitachi SH-4 @ 200 MHz

  - Features: 32-bit SIMD @ 200 MHz, floating-point unit, graphic functions

  - Performance: 360 MIPS and 1.4 GFLOPS

- GPU: NEC-VideoLogic PowerVR 2 (PVR2DC/CLX2) @ 100 MHz[166]

  - Texture mapping: Bump mapping, mipmapping,[167] environment mapping, texture compression,[168] multi-texturing,[169] perspective correction[166]

  - Filtering: Point filtering,[166] bilinear filtering,[168] trilinear filtering, anisotropic filtering[166]

  - Anti-aliasing: Super-sampling anti-aliasing (SSAA),[166] full-scene anti-aliasing (FSAA)[169]

- Alpha blending: 256 levels of transparency,[166] multi-pass blending,[169] translucency sorting[169]

- Shading: Perspective-correct ARGB Gouraud shading,[169] shadows[166]

- Rendering: ROP (render output unit), tiled rendering, 32-bit floating-point Z-buffering, 32-bit floating-point hidden surface removal,[169] 256 fog effects,[166] per-pixel table fog[169]

- Other capabilities: Quad polygons, triangle polygons, GMV (general modifier volumes)[166]

- Sound engine: Yamaha AICA Super Intelligent Sound Processor @ 45 MHz[168]

  - Internal CPU: 32-bit ARM7 RISC CPU @ 45 MHz

  - CPU performance: 40 MIPS[155]

  - PCM/ADPCM: 16-bit depth, 48 kHz sampling rate (DVD quality), 64 channels[166]

  - Other features: DSP, sound synthesizer

- Operating system: Windows CE[166] (with DirectX 6.0, Direct3D, and OpenGL)

- RAM: 56 MB (64 MB with GD-ROM)

  - Main RAM: 32 MB

  - VRAM: 16 MB[167] (unified framebuffer and texture memory)[169][170]

  - Sound memory: 8 MB

  - DIMM: 8 MB DRAM (GD-ROM variants only)[170]

- Storage media:

  - ROM board: Up to 172 MB

  - Disc storage: GD-ROM (1 GB) drive @ 12× speed[166]

- Display resolution: VGA,[171] 320×240 to 800×608 pixels,[172] progressive scan

- Color depth: 32-bit[169] ARGB,[166] 16,777,216 colors (24-bit color)[167] with 8-bit (256 levels) alpha blending,[166][169] YUV and RGB color spaces, color key overlay[169]

- Polygon performance: 7 million textured polygons/sec (with shadows,[173] lighting[168] and trilinear filtering[174]) to 10 million polygons/sec (with lighting)[168][175]

- Rendering fillrate: 500 million pixels/sec[176] (with transparent polygons) to over 3.2 billion pixels/sec (with opaque polygons)[168]

- Texture fillrate: 100 million texels/sec (up to 1.6 billion texels/sec in Multiboard)

**NAOMI Multiboard specifications** Sega NAOMI Multiboard included the following upgrades in 1999:[*][165]

- CPU: 2× to 16× Hitachi SH-4 @ 200 MHz

    - Performance: 720 to 5760 MIPS, 2.8 to 22.4 GFLOPS

- GPU: 2× to 16× NEC-VideoLogic PowerVR 2 (PVR2DC/CLX2) @ 100 MHz

- Sound engine: 2× to 16× Yamaha AICA Super Intelligent Sound Processor @ 45 MHz

    - Internal CPU: 2× to 16× 32-bit ARM7 RISC CPU @ 45 MHz

    - CPU performance: 80 to 640 MIPS

    - PCM/ADPCM: 128 to 1024 channels

- RAM: 112 to 896 MB (128 to 1024 MB with GD-ROM)

    - Main RAM: 64 to 512 MB

    - VRAM: 32 to 256 MB

    - Sound memory: 16 to 128 MB

- Storage media:

    - ROM boards: 344 to 2752 MB

    - Disc storage: 2 to 16 GD-ROM drives

- Display resolution: 3-monitor widescreen VGA,[*][165] 960×240 to 2400×608 pixels, progressive scan

- Polygon performance: 14 to 112 million textured polygons/sec (with lighting and trilinear filtering), or 20 to 160 million polygons/sec

- Rendering fillrate: 1 to 8 billion pixels/sec (with transparent polygons), 6.4 to 51.2 billion pixels/sec (with opaque polygons)

- Texture fillrate: 200 million to 1.6 billion texels/sec

## Sega Hikaru

An evolution of the NAOMI hardware with superior graphics capabilities, the Hikaru was used for a handful of deluxe dedicated-cabinet games, beginning with 1999's *Brave Fire Fighters,* in which the flame and water effects were largely a showpiece for the hardware. The Hikaru hardware was the first arcade platform capable of effective Phong shading.

According to Sega in 1999: *"Brave Firefighters utilizes a slightly modified Naomi Hardware system called Hikaru. Hikaru incorporates a custom Sega graphics chip and possesses larger memory capacity than standard Naomi systems. "These modifications were necessary because in Brave Firefighters, our engineers were faced with the daunting challenge of creating 3d images of flames and sprayed water,"* stated Sega's Vice President of Sales and Marketing, Barbara Joyiens. *"If you stop and think about it, both have an almost infinite number of shapes, sizes, colors, levels of opaqueness, shadings and shadows. And, when you combine the two by simulating the spraying of water on a flame, you create an entirely different set of challenges for our game designers and engineers to overcome; challenges that would be extremely difficult, if not impossible to overcome utilizing existing 3D computers. Hikaru has the horsepower to handle these demanding graphic challenges with clarity, depth and precision."*[*][177] In addition, the Hikaru also uses two Hitachi SH-4 CPU's, two Yamaha AICA sound engines,[*][178] a Motorola 68000 network CPU, and two PowerVR2 GPU's.[*][179]

Since it was comparatively expensive to produce, and most games did not necessarily need Hikaru's extended graphics capabilities, Sega soon abandoned the system in favor of continued NAOMI and NAOMI 2 development.

## Hikaru specifications

- Main CPU: 2× Hitachi SH-4 @ 200 MHz[*][178]

    - Features: 2× 128-bit SIMD @ 200 MHz, 2× floating-point units, graphic functions

    - Performance: 720 MIPS and 2.8 GFLOPS

- Network CPU: Motorola 68000[*][179]

- Sound engine: 2× Yamaha AICA Super Intelligent Sound Processor @ 45 MHz[*][178]

    - Internal CPU: 2× 32-bit ARM7 RISC CPU @ 45 MHz

    - CPU performance: 34 MIPS (2× 17 MIPS)[*][168]

    - PCM/ADPCM: 16-bit depth, 48 kHz sampling rate (DVD quality),[*][166] 128 channels[*][178]

    - Other features: DSP, sound synthesizer

- GPU: 2×[*][179] NEC-VideoLogic PowerVR 2 (PVR2DC/CLX2) @ 100 MHz[*][166]

    - Texture mapping: Bump mapping, mipmapping,[*][167] environment mapping, texture compression,[*][168] multi-texturing,[*][169] perspective correction[*][166]

- Filtering: Point filtering,[*][166] bilinear filtering,[*][168] trilinear filtering, anisotropic filtering[*][166]

- Anti-aliasing: Super-sampling anti-aliasing (SSAA),[*][166] full-scene anti-aliasing (FSAA)[*][169]

- Alpha blending: 256 levels of transparency,[*][166] multi-pass blending,[*][169] translucency sorting[*][169]

- Shading: Perspective-correct ARGB Gouraud shading,[*][169] shadows[*][166]

- Rendering: ROP (render output unit), tiled rendering, 32-bit floating-point Z-buffering, 32-bit floating-point hidden surface removal,[*][169] 256 fog effects,[*][166] per-pixel table fog[*][169]

- Other capabilities: Quad polygons, triangle polygons, GMV (general modifier volumes)[*][166]

- T&L Graphics Engine: Sega Custom 3D[*][177][*][178]

  - Lighting: Horizontal, spot, 1024 lights per scene, 4 lights per polygon, 8 window surfaces[*][178]

  - Shading: Phong shading, shadow[*][178]

  - Rendering: Fog, depth queueing

  - Other effects: Stencil, motion blur,[*][178] particle effects, fire effects, water effects[*][177]

  - Other capabilities: 2 bitmap layers, calendar

- Operating system: Windows CE[*][166] (with DirectX 6.0, Direct3D, and OpenGL)

- RAM: 100 MB

  - Main RAM: 64 MB[*][178]

  - VRAM: 28 MB[*][178]

  - Sound SDRAM:[*][179] 8 MB[*][178]

  - Network SRAM: 32 KB[*][179]

- Storage media: ROM Board, up 352 MB

- Color depth: 32-bit[*][169] ARGB,[*][166] 16,777,216 colors (24-bit color)[*][167] with 8-bit (256 levels) alpha blending,[*][166][*][169] YUV and RGB color spaces, color key overlay[*][169]

- Display resolution: 31 kHz horizontal sync,[*][178] 60 Hz refresh rate,[*][179] VGA,[*][171] progressive scan

  - Single monitor: 496×384[*][178] to 800×608 pixels[*][172]

  - Dual monitor:[*][178] 992×768 to 1600×608 pixels

- Polygon performance:

  - With Phong shading, 4 lights per polygon, shadows, trilinear filtering, motion blur and all other effects: 4 million textured polygons/sec (2 million per GPU[*][178])

  - With lighting, shadows, and trilinear filtering: 14 million textured polygons/sec (7 million per GPU[*][173])

  - With lighting: 20 million polygons/sec (10 million per CPU/GPU[*][168][*][175])

- Fillrate:

  - Rendering: 1 billion pixels/sec (with transparent polygons) to over 6.4 billion pixels/sec (with opaque polygons)

  - Textures: 200 million texels/sec

- Extensions: communication, 4-channel surround audio, PCI, MIDI, RS-232C

- Connection: JAMMA Video compliant

**Sega NAOMI 2**

In 2000, Sega debuted the *NAOMI 2* arcade system board at JAMMA, an upgrade and a sequel of the original NAOMI with better graphics capability.

NAOMI 2's graphics-assembly contains two PowerVR CLX2 GPUs, a PowerVR Elan chip for geometry transformation and lighting effects, and 2X the graphics memory for each CLX2 chip. (Each CLX2 has its own 32MB bank, as the CLX2s cannot share graphics RAM). Due to architectural similarities and a "bypass" feature in the Elan device, the NAOMI 2 is also able to play NAOMI games without modification.[*][180][*][181][*][182]

With the NAOMI 2, Sega brought back the GD-ROM drive. For both NAOMI and NAOMI 2, the GD-ROM setup was offered as an optional combination of daughterboard expansion known as the DIMM Board, and the GD-ROM drive itself. The DIMM board contained enough RAM to allow an entire game to be loaded into memory at start up, allowing the drive to shut down after the game has loaded. This heavily reduces load times during the game, and saves on drive wear and tear.

## 4.2.16 Triforce

The *Triforce* is an arcade system board developed jointly by Namco, Sega, and Nintendo, with the first games appearing in 2002. The name "Triforce" is a reference to

Nintendo's *The Legend of Zelda* series of games, and symbolized the three companies' involvement in the project. The system hardware is based on the Nintendo GameCube with several differences, like provisions for add-ons such as Sega's GD-ROM system and upgradeable RAM modules. The Triforce was initially believed to have twice as much 1T-SRAM as the Nintendo GameCube (48MB instead of 24MB), but this was disproven by a teardown analysis of a Triforce board.[183]

A few versions of the Triforce exist. The first two are the Type-1 and Type-3 units, the former using an external DIMM board (same as used on the Naomi and Naomi 2) while the latter integrates this component inside the metal casing. A custom Namco version exists which only accepts custom NAND Flash based cartridges, which has a different Media board and supposedly different baseboard.[184] These boards use the same metal case design as the Type-3 Triforce.

### Triforce specifications

- Main CPU: IBM PowerPC "Gekko" @ 486 MHz

- Graphics: Custom ATI/Nintendo "Flipper" @ 162 MHz.

- Color: 24-bit color (24-bit z-buffer)

- Hardware features: Fog, subpixel anti-aliasing, 8 hardware lights, alpha blending, virtual texture design, multi-texturing, bump mapping, environment mapping, mipmapping, bilinear filtering, trilinear filtering, anisotropic filtering, real-time hardware texture decompression (S3TC), real-time decompression of display list, embedded framebuffer, 1 MB embedded texture cache, 3-line deflickering filter.

- Sound DSP: Custom Macronix 16-bit DSP @ 81 MHz

- Main RAM: Main memory 24 MB of MoSys 1T-SRAM, approximately 10 ns sustainable latency.[185][186]

### Porting

In 2012, a homebrew application was released for the Nintendo Wii that enabled this GameCube-derived console to run *Mario Kart Arcade GP*, *Mario Kart Arcade GP 2*, *F-Zero AX* and *Virtua Striker 4 Ver.2006* (see the list of games below). The coder stated that support for other games and additional features are possible. The homebrew application Nintendont, designed for running GameCube games off USB drives, is also capable of running games designed for the Triforce hardware.

## 4.2.17   Sega Chihiro

The *Sega Chihiro* system is a Sega arcade system board based on the architecture of the Xbox. The 733 MHz Intel Pentium III CPU and the Nvidia XChip graphics processor are common to both, but the Chihiro has a different MCPX chip with unique bootloader keys. The main system memory, at 128 MB, is twice that of a retail Xbox. In addition to this memory, the Chihiro also has additional RAM used for media storage - this was initially 512 MB but is upgradable to 1 GB. When the system is booted, the required files are copied from the GD-ROM to the RAM on the media board.

Because the Chihiro and Xbox share the same hardware architecture, porting from the Chihiro is theoretically easier than porting from a different arcade platform. In practice, there are a number of challenges - the first being that the half-size main memory restricts the size of your working set and the second being that fetching assets from Xbox DVD drive is orders of magnitude slower than fetching them from the 512MB/1GB of RAM on the media board. These challenges are not insurmountable, though - for example, the Xbox release of *OutRun 2* was able to retain the look and feel of the original arcade version.

### Chihiro specifications

- CPU: Pentium III @ 733 MHz, 133 MHz FSB

- System RAM: 128 MB soldered on main PCB

- Media RAM: 512 MB upgradable to 1 GB (DIMM on Media board)

- GPU: Nvidia XChip @ 200 MHz, (derived from GeForce 3), featuring programmable pixel and vertex shaders, hardware T&L, Quincunx FSAA, anisotropic filtering, bump mapping

- Sound: Cirrus Logic CS4630 Stream Processor, Nvidia nForce with 5.1 Dolby Digital decoding

- Media: GD-ROM[187]

## 4.2.18   Sega Lindbergh

The *Sega Lindbergh* arcade system board is an embedded PC running MontaVista Linux (The Lindbergh Blue system used Windows Embedded instead). Sega had initially planned to use Microsoft's Xbox 360 as the basis for the arcade board, but instead opted for an architecture based on standard PC hardware.

According to Sega-AM2 president Hiroshi Kataoka, porting Lindbergh titles (such as Virtua Fighter 5) to Sony's

PlayStation 3 is generally easier than porting to Xbox 360, because the Lindbergh and PS3 use a GPU designed by the same company, Nvidia.*[188]

**Lindbergh specifications**

- CPU: Pentium 4 HT 3.0E (3.0 GHz, 1 MB L2 Cache, Hyper-Threading, 800 MHz FSB)

- RAM: 184-pin DDR SDRAM PC3200 (400 MHz) 512 MB × 2 (Dual)

- GPU: Nvidia GeForce 6800 AGP (NV40), 256 Bit GDDR3 256 MB, compatible with Vertex Shader 3.0 & Pixel Shader 3.0

- Sound: 64 channel, 5.1 ch S/PDIF

- LAN: On board, 10/100/1000 BASE-TX. JVS I/O Connector

- Serial: 2 Channel (can switch one channel between 232C and 422)

- Other: USB port x 4, high-definition output (DVI and VGA out), S-Video out, DVD Drive Support, Sega ALL.NET online support

- Operating System: MontaVista Linux*[189]

- Protection : High spec original security module.

The Sega Lindbergh standard universal sit-down cabinet uses a 1360 × 768 WXGA LCD display.

Aside from the standard Lindbergh system (Lindbergh Yellow), Sega developed a Lindbergh Red which includes the GeForce 7600gs and Lindbergh Blue system, which have different specifications.

The Lindbergh has been superseded by the Ring series (RingEdge and RingWide), so there will be no new arcade games developed for this system. The last game to run on Lindbergh was *MJ4 Evolution.*\*[190]

### 4.2.19   Sega Europa-R

The *Sega Europa-R* is an arcade system board developed by Sega Amusements Europe.

Sega chose a PC-based design for this arcade board. This arcade board currently only runs two games, Sega Rally 3 and Race Driver: GRID (Stylized as simply GRID).

**Europa-R specifications**

- CPU: Intel Pentium D 945 (3.4 GHz, dual-core)

- RAM: 8 GB (2x 4 GB modules)

- GPU: Nvidia GeForce 8800

- Other: Compatible HDTV (High Definition), DVD drive support, Sega ALL.NET online support

- Protection: High spec original security module.

### 4.2.20   Sega Ring series

The Ring series of arcade machines are also based on PC architecture. Initially announced models include *RingEdge* and *RingWide*. The 2 pieces of hardware have Microsoft Windows Embedded Standard 2009 as their operating system, mainly so other third-party companies would find it easier to produce games for the system.

**RingEdge**

The RingEdge is the main console of the Ring Series. It has better graphics and larger storage than the RingWide. It sports a better graphics card than the Lindbergh system, allowing for a higher performance graphically, all while costing less to produce. The use of an Intel Pentium Dual-Core (1.8 GHz per core) processor delivers better performance than Lindbergh's Pentium 4 (3.0 GHz) processor. A solid-state drive greatly reduces wear-and-tear due to a lack of moving parts, and also has much higher transfer rates than a hard disc drive, leading to better performance and loading times. The Ringedge also supports 3D game capability.

**RingEdge specifications**

- CPU: Intel Pentium Dual-Core E2160 (1.8 GHz)

- RAM: 1 GB DDR2 SDRAM (PC-6400)

- GPU: Nvidia GeForce 8800GS with 384 MB GDDR3 SDRAM (Shader Model 4.0)

- Output: 2 DVI ports

- Storage: 32 GB SSD

- Networking: Gigabit Ethernet (1000BASE-T)

- OS: Microsoft Windows Embedded Standard 2009

- Sound: 5.1 channel HD Audio

- Other: 3 USB ports, Sega ALL.NET online support

### RingWide

The RingWide is more basic than the RingEdge, and only has 8 GB (CompactFlash) of storage, while RingEdge has a four times larger storage (because of the use of the RAM Drive and SSD). The RingWide will be used to run games that are less graphics-intensive and that require less high-end specifications in order to cut down costs. Sega also appears poised to be designing a streaming hybrid for use with household TVs, similar to OnLive from the system's hardware as evident from this patent issued by them on November 17, 2009.*[191]

### RingWide specifications

- CPU: Intel Celeron 440 (2.0 GHz)

- RAM: 1 GB DDR2 SDRAM (PC-5300)

- GPU: ATI Radeon HD 2xxx with 128 MB GDDR3 SDRAM (Shader Model 4.0)

- Output: 1 DVI port

- Storage: 8 GB CompactFlash

- Networking: Gigabit Ethernet (1000BASE-T)

- OS: Microsoft Windows Embedded Standard 2009

- Other: 2 USB ports, 5.1 channel HD Audio, Sega ALL.NET online

### RingEdge 2

The successor to RingEdge,

### RingEdge 2 specifications

- CPU: Intel Core i3 540 3.07 GHz

- GPU: Nvidia GeForce GT 545 GDDR5 with 1GB GDDR5 memory (Direct3D 11.1/OpenGL 4.3)

- RAM: 4GB DDR3 SDRAM (PC3-12800)

- Sound: 5.1 channel HD Audio

- Output: DVI-I, DVI-D, twin display output

- Connectors: JVS I/O connector, 4 channel serial port, 4× USB 3.0, 2× CAN bus

- Networking: Gigabit Ethernet (1000BASE-T)

- Storage: 32GB TDK GBDISK RS3 SSD

- OS: Windows Embedded Standard 2012(Windows 7)

- Media: DVD or USB storage, network delivery (Sega ALL.NET)

## 4.2.21  Sega Nu

Released in Japan in November 2013. Nu is based on a mid-range PC running Windows 8.

### Nu specifications

- CPU: Intel Core i3−3220 3.30 GHz

- RAM: 4 GB DDR3 SDRAM (PC3-12800)

- GPU: Nvidia GeForce GTX 650 Ti with 1 GB GDDR5 memory (Direct3D 11.1/OpenGL 4.3)

- Sound: 5.1 channel HD Audio

- Output: DVI-I, DVI-D, twin display output

- Connectors: JVS I/O connector, 4 channel serial port, 4× USB 3.0, 2× CAN bus

- Networking: Gigabit Ethernet (1000BASE-T)

- Storage: SATA SSD 64 GB, HDD 500 GB

- OS: Microsoft Windows Embedded 8 Standard

- Media: DVD or USB storage, network delivery (Sega ALL.NET)

## 4.2.22  Technical details

The "suicide battery" (System 18, System 16 and others) generally refers to an arrangement by which encryption keys or other vital data are stored in SRAM powered by a battery. When the battery dies, the PCB is rendered permanently inoperable, in the sense that there is no way to reprogram the RAM from within the PCB itself —hence the term "suicide".

## 4.2.23  See also

- List of Sega arcade games

- List of Sega video game franchises

- R-360

- List of game engines

## 4.2.24 References

[1] http://www.system16.com/hardware.php?id=684

[2] https://github.com/mamedev/mame/tree/master/src/mame/drivers/blockade.c

[3] http://web.archive.org/web/20120424231244/http://www.depi.itch.edu.mx/apacheco/asm/Intel_cpus.htm

[4] http://www.system16.com/hardware.php?id=685

[5] https://github.com/mamedev/mame/tree/master/src/mame/drivers/vicdual.c

[6] http://www.drolez.com/retro/

[7] http://www.arcade-history.com/?n=depthcharge&page=detail&id=622

[8] https://github.com/mamedev/mame/tree/master/src/mame/video/vicdual.c

[9] http://www.system16.com/hardware.php?id=688

[10] https://github.com/mamedev/mame/tree/master/src/mame/drivers/galaxian.c

[11] http://books.google.co.uk/books?id=lB4PAwAAQBAJ&pg=PA181

[12] https://github.com/mamedev/mame/tree/master/src/mame/drivers/galdrvr.c

[13] https://github.com/mamedev/mame/tree/master/src/mame/video/galaxian.c

[14] http://www.system16.com/hardware.php?id=513

[15] https://github.com/mamedev/mame/tree/master/src/mame/includes/galaxian.h

[16] http://www.arcade-museum.com/game_detail.php?game_id=7885

[17] http://www.cs.columbia.edu/~{}%20sedwards/classes/2011/4840/reports/Galaxian.pdf

[18] https://github.com/mamedev/mame/tree/master/src/emu/sound/sn76496.c

[19] https://github.com/mamedev/mame/tree/master/src/mame/drivers/suprloco.c

[20] http://imame4all.googlecode.com/svn-history/r146/Reloaded/trunk/src/mame/video/segaic16.c

[21] https://github.com/mamedev/mame/tree/master/src/mame/video/suprloco.c

[22] https://github.com/mamedev/mame/tree/master/src/mame/video/bankp.c

[23] https://github.com/mamedev/mame/tree/master/src/mame/drivers/bankp.c

[24] "Sega G80 Hardware Reference". Archived from the original on 2012-02-19.

[25] https://github.com/mamedev/mame/tree/master/src/mame/drivers/segag80r.c

[26] "SEGA G80 Raster Hardware".

[27] http://www.system16.com/hardware.php?id=686

[28] https://archive.org/stream/bitsavers_inteldataSngleComponent8BitMicrocomputerDataSheet1_846962/8048_8035_HMOS_Single_Component_8-Bit_Microcomputer_DataSheet_1980

[29] https://github.com/mamedev/mame/tree/master/src/mame/audio/segag80r.c

[30] http://web.archive.org/web/20130104202105/http://mamedev.org/source/src/mame/video/segag80r.c.html

[31] https://web.archive.org/web/20130104202114/http://mamedev.org/source/src/mame/video/segag80v.c.html

[32] http://www.system16.com/hardware.php?id=690

[33] https://github.com/mamedev/mame/tree/master/src/mame/video/turbo.c

[34] http://flyers.arcade-museum.com/?page=thumbs&db=videodb&id=1106

[35] http://www.system16.com/files/manuals/subroc3d.pdf

[36] http://www.solvalou.com/subpage/arcade_reviews/173/479/subroc-3d_review.html

[37] https://github.com/mamedev/mame/tree/master/src/mame/drivers/turbo.c

[38] https://github.com/mamedev/mame/tree/master/src/mame/includes/turbo.h

[39] http://www.system16.com/hardware.php?id=689

[40] https://github.com/mamedev/mame/tree/master/src/mame/drivers/zaxxon.c

[41] http://www.vasulka.org/archive/Writings/VideogameImpact.pdf

[42] https://github.com/mamedev/mame/tree/master/src/mame/video/zaxxon.c

[43] https://github.com/mamedev/mame/tree/master/src/mame/drivers/segald.c

[44] http://www.system16.com/hardware.php?id=691

[45] http://www.blam1.com/LaserDisc/FAQ/

[46] "Sega System 1 game and hardware information".

[47] https://github.com/mamedev/mame/tree/master/src/mame/drivers/system1.c

[48]  https://github.com/mamedev/mame/tree/master/src/mame/video/system1.c

[49]  IGN Presents the History of SEGA: World War, IGN

[50]  http://www.system16.com/hardware.php?id=695

[51]  "System 16 (A version) at System 16 - The Arcade Museum".

[52]  "System 16 (B version) at System 16 - The Arcade Museum".

[53]  http://pdf.datasheetcatalog.com/datasheet/Intel/mXuwzsy.pdf

[54]  http://cgfm2.emuviews.com/txt/p16tech.txt

[55]  "Sega AGES Vol.33 FANTASY ZONE COMPLETE COLLECTION" Sega Release on 2008-09-11 in Japan.

[56]  https://github.com/mamedev/mame/tree/master/src/mame/drivers/segas16a.c

[57]  Sega's 16-bit arcade color palette: 15-bit RGB high color depth (32,768 colors) and 1-bit shadow & highlight that triples up to 98,304 colors.

[58]  http://cgfm2.emuviews.com/txt/s16tech.txt

[59]  http://pdf.datasheetcatalog.com/datasheets2/18/188119_1.pdf

[60]  https://github.com/mamedev/mame/tree/master/src/emu/sound/upd7759.c

[61]  "Sega Museum - Sega System 24 Hardware".

[62]  https://github.com/mamedev/mame/tree/master/src/mame/drivers/segas24.c

[63]  http://cgfm2.emuviews.com/fdconv.php

[64]  http://cgfm2.emuviews.com/txt/s24tech.txt

[65]  http://cgfm2.emuviews.com/new/s24hw.txt

[66]  "System 18 at System 16 - The Arcade Museum".

[67]  https://github.com/mamedev/mame/tree/master/src/mame/drivers/segas18.c

[68]  https://github.com/mamedev/mame/tree/master/src/mame/drivers/kyugo.c

[69]  http://www.system16.com/hardware.php?id=861

[70]  https://github.com/mamedev/mame/tree/master/src/mame/video/kyugo.c

[71]  http://retro.ign.com/articles/974/974695p3.html

[72]  http://www.extentofthejam.com/pseudo/

[73]  http://web.archive.org/web/20131113174154/http://www.1up.com/features/disappearance-suzuki-part-1?pager.offset=2

[74]  https://github.com/mamedev/mame/tree/master/src/mame/drivers/segahang.c

[75]  http://www.system16.com/hardware.php?id=696

[76]  https://github.com/mamedev/mame/tree/master/src/mame/drivers/segaorun.c

[77]  http://cgfm2.emuviews.com/txt/loftech.txt

[78]  http://www.theregister.co.uk/2014/02/18/antique_code_show_space_harrier/

[79]  http://dallasdoan.com/misc/eBooks/GameDesignEbooksColletion/VideoGameTheory.pdf

[80]  http://www.system16.com/hardware.php?id=697

[81]  http://imame4all.googlecode.com/svn-history/r146/Reloaded/trunk/src/mame/drivers/segaorun.c

[82]  http://www.hardcoregaming101.net/outrun/outrun.htm

[83]  http://www.mikesarcade.com/cgi-bin/spies.cgi?action=url&type=info&page=outrunFAQ.txt

[84]  http://www.theregister.co.uk/2013/12/18/antique_code_show_sega_out_run/

[85]  http://www.coinop.org/kb_dl.aspx/KB/faqs/faq-sega%20outrun.html

[86]  https://github.com/mamedev/mame/tree/master/src/mame/drivers/segaxbd.c

[87]  "X Board at system16.com".

[88]  http://www.hardcoregaming101.net/afterburner/afterburner.htm

[89]  "Y Board at system16.com".

[90]  https://github.com/mamedev/mame/tree/master/src/mame/drivers/segaybd.c

[91]  http://web.archive.org/web/20130104202220/http://mamedev.org/source/src/mame/video/segaybd.c.html

[92]  João Diniz-Sanches, ed. (November 2003). "Closer to the Heart". Edge (Bath: Future Publishing) (129): 87. Aside from the three different base unites, the internal workings of the Mega Drive found itself in a choice of guises, including... Megatech (an eight-way arcade cabinet that played Mega Drive games), Mega-play (a JAMMA-compatible arcade cabinet with Mega Drive software on proprietary boards, similar to SNK's MVS)...

[93]  "Mega-Tech".

[94]  "Mega Play".

[95]  "Sega System C-2 information at System 16".

[96]  https://github.com/mamedev/mame/tree/master/src/mame/drivers/segac2.c

[97] http://datasheet.eeworld.com.cn/pdf/NEC/71665_UPD7759GC-3BH.pdf

[98] "System 32 hardware information and game screen shots".

[99] "Model 1 at system16.com".

[100] http://ipsj.ixsq.nii.ac.jp/ej/?action=pages_view_main&active_action=repository_view_main_item_detail&item_id=59745&item_no=1&page_id=13&block_id=8

[101] http://multimedia.cx/NEC_V60pgmRef.pdf

[102] https://github.com/mamedev/mame/tree/master/src/mame/drivers/segas32.c

[103] http://pdf1.alldatasheet.com/datasheet-pdf/view/122826/HITACHI/HM53461ZP-12.html

[104] http://www.datasheetarchive.com/dlmain/Datasheets-22/DSA-431594.pdf

[105] http://tech.quarterarcade.com/tech/MAME/src/system32.c.html.aspx?g=1395

[106] http://web.archive.org/web/20130104202207/http://mamedev.org/source/src/mame/video/segas32.c.html

[107] http://www.retroroms.net/modules/news/index.php?storytopic=1&start=60

[108] http://www.retrogames.com/012003.html

[109] http://books.google.co.uk/books?id=DbFxAgAAQBAJ&pg=PA63

[110] http://www.gamesetwatch.com/2007/03/column_arcade_obscurities_sega.php

[111] http://www.system16.com/hardware.php?id=710

[112] http://archive.computerhistory.org/resources/access/text/2013/04/102723432-05-01-acc.pdf

[113] https://github.com/mamedev/mame/tree/master/src/mame/drivers/model1.c

[114] http://members.iinet.net.au/~{}lantra9jp1/gurudumps/m2status/index.html

[115] http://web.archive.org/web/20130104200817/http://mamedev.org/source/src/mame/video/model1.c.html

[116] http://www.consoledatabase.com/faq/segasaturn/segasaturnfaq.txt

[117] http://vintage3d.org/history.php#sthash.Wlg79A2P.dpbs

[118] http://uk.ign.com/articles/2009/04/21/ign-presents-the-history-of-sega?page=8

[119] http://pdf.datasheetcatalog.com/datasheet/Intel/mXqwttu.pdf

[120] "Model 2 Hardware (Sega)". System 16. Retrieved 2011-04-18.

[121] http://www.system16.com/hardware.php?id=714

[122] http://www.arcade-history.com/?n=virtua-fighter-2&page=detail&id=3328

[123] http://www.system16.com/hardware.php?id=715

[124] http://mamereviews.hubmed.org/game/vstrikro

[125] http://www.system16.com/hardware.php?id=716

[126] http://datasheets.chipdb.org/Intel/80960/PRODBREF/27223303.PDF

[127] http://web.archive.org/web/20130104200822/http://mamedev.org/source/src/mame/video/model2.c.html

[128] http://www.analog.com/static/imported-files/data_sheets/ADSP-21060_21060L_21062_21062L_21060C_21060LC.pdf

[129] https://github.com/mamedev/mame/tree/master/src/mame/drivers/model2.c

[130] http://www.hotchips.org/wp-content/uploads/hc_archives/hc07/3_Tue/HC7.S5/HC7.5.1.pdf

[131] http://www.hotchips.org/wp-content/uploads/hc_archives/hc07/3_Tue/HC7.S5/HC7.5.1.pdf#page=4

[132] http://koti.kapsi.fi/~{}antime/sega/files/ST-077-R2-052594.pdf

[133] http://www.gamezero.com/team-0/whats_new/past/news9504.html

[134] http://www.fujitsu.com/downloads/MAG/vol33-2/paper08.pdf

[135] http://www.hotchips.org/wp-content/uploads/hc_archives/hc07/3_Tue/HC7.S5/HC7.5.1.pdf#page=8

[136] http://www.hotchips.org/wp-content/uploads/hc_archives/hc07/3_Tue/HC7.S5/HC7.5.1.pdf#page=12

[137] "News: Virtua Fighter 3". *Computer and Video Games* (174): 10–1. May 1996.

[138] "Model 3 Step 1.0 at system16.com".

[139] https://github.com/mamedev/mame/tree/master/src/mame/drivers/model3.c

[140] http://www.system16.com/hardware.php?id=718

[141] http://www.system16.com/hardware.php?id=719

[142] http://www.system16.com/hardware.php?id=720

[143] http://web.stanford.edu/dept/HPS/TimLenoir/MilitaryEntertainmentComplex.htm

[144] http://www.fermimn.gov.it/inform/materiali/evarchi/ motorola/603e_fs.pdf

[145] http://www.segatech.com/archives/january1998.html

[146] http://www.supermodel3.com/About.html

[147] http://web.archive.org/web/20130104200833/http: //mamedev.org/source/src/mame/video/model3.c.html

[148] http://www.vgamuseum.info/index.php/glossary/ Glossary-1/3/3D-RAM-20/

[149] http://ieeexplore.ieee.org/xpl/login.jsp?tp=&arnumber= 535565&url=http%3A%2F%2Fieeexplore.ieee.org% 2Fxpls%2Fabs_all.jsp%3Farnumber%3D535565

[150] http://www.gamespot.com/articles/ more-on-segas-model-3-step-2-board/1100-2462493/

[151] http://www.datasheets360.com/part/detail/ m5m410092fp-15/$-$7080454212028813194/

[152] http://www.datasheetarchive.com/dlmain/Datasheets-13/ DSA-247062.pdf

[153] http://www.datasheets360.com/part/detail/ m5m4v4169tp-20/1857732301728935217/

[154] http://www.system16.com/hardware.php?id=711&page=3

[155] http://www.segatech.com/technical/saturnspecs/

[156] "Sega Titan Video at System 16 - The Arcade Museum".

[157] http://web.archive.org/web/20130104202915/http: //mamedev.org/source/src/mame/video/stvvdp1.c.html

[158] http://www.sega-saturn.com/saturn/other/satspecs.htm

[159] http://koti.kapsi.fi/~{}antime/sega/files/ ST-013-R3-061694.pdf#page=81

[160] http://koti.kapsi.fi/~{}antime/sega/files/ ST-013-R3-061694.pdf#page=119

[161] http://koti.kapsi.fi/~{}antime/sega/files/ ST-013-R3-061694.pdf#page=18

[162] http://koti.kapsi.fi/~{}antime/sega/files/ ST-013-R3-061694.pdf#page=75

[163] http://koti.kapsi.fi/~{}antime/sega/files/ ST-013-R3-061694.pdf#page=24

[164] http://web.archive.org/web/20130104202921/http: //mamedev.org/source/src/mame/video/stvvdp2.c.html

[165] http://www.system16.com/hardware.php?id=906

[166] http://segatech.com/technical/overview/index.html

[167] "NAOMI technical overview".

[168] http://web.archive.org/web/20000823204755/http: //computer.org/micro/articles/dreamcast_2.htm

[169] http://web.archive.org/web/20070811102018/http: //www3.sharkyextreme.com/hardware/reviews/video/ neon250/2.shtml

[170] https://github.com/mamedev/mame/tree/master/src/ mame/drivers/naomi.c

[171] http://wiki.arcadeotaku.com/w/Sega_Naomi_Universal

[172] http://cadcdev.sourceforge.net/docs/kos-current/video_ 8h_source.html

[173] http://segatech.com/technical/gpu/index.html

[174] http://books.google.co.uk/books?id=wZnpAgAAQBAJ& pg=PA277

[175] http://segatech.com/technical/cpu/index.html

[176] http://www.system16.com/hardware.php?id=721

[177] http://www.goodcowfilms.com/farm/games/news-archive/ Sega%20Confirms%20Hikaru%20DOES%20Exist....htm

[178] http://www.system16.com/hardware.php?id=724

[179] https://github.com/mamedev/mame/tree/master/src/ mame/drivers/hikaru.c

[180] "NAOMI 2 GD-ROM Hardware". System 16. Retrieved 2006-08-02.

[181] "NAOMI 2 Hardware". System 16. Retrieved 2006-08-02.

[182] "NAOMI 2 Satellite Terminal hardware". System 16. Retrieved 2006-08-02.

[183] "Triforce hardware analysis".

[184] "Triforce Baseboard".

[185] "System16 - Sega Triforce".

[186] "Triforce Hardware".

[187] "Chihiro hardware specifications and known games".

[188] "VF5 port to PS3".

[189] "Pictures of actual *Virtua Tennis 3* bootup exposing Linux".

[190] "セガネットワーク対戦麻雀 MJ4 Evolution 公式サイト". Sega-mj.com. Retrieved 2011-04-18.

[191] "Network game system,and game terminal device and storage medium".

## 4.2.25   External links

- Phantom's Arcade World

- Sega list @ PCBdB*

# Chapter 5

# Text and image sources, contributors, and licenses

## 5.1 Text

- **Sega Saturn** *Source:* https://en.wikipedia.org/wiki/Sega_Saturn?oldid=680466910 *Contributors:* The Epopt, Derek Ross, Tarquin, Jzcool, Ffaker, Infrogmation, Michael Hardy, Bewildebeast, Brtkrbzhnv, Minesweeper, Ahoerstemeier, Typhoon, Schneelocke, Ideyal, Crissov, Blargg, WhisperToMe, IceKarma, Furrykef, K1Bond007, Shizhao, Topbanana, Bloodshedder, MD87, Robbot, Meelar, Mushroom, Alexwcovington, Misterkillboy, Samusfan80, FriedMilk, Xinoph, Tom-, Saaga, Bobblewik, Golbez, Mooquackwooftweetmeow, Wmahan, Chowbok, LiDaobing, UgenBot, Kusunose, Mamizou, Bumm13, Sam Hocevar, Srittau, Shadowlink1014, Chmod007, Damieng, Lacrimosus, Neonchameleon, Juan Ponderas, Slady, Econrad, Rich Farmbrough, Guanabot, Vague Rant, Pixel8, Smyth, Shadow Hog, Indrian, Bender235, ESkog, Nekochan, Ht1848, Sockatume, Juppiter, Sietse Snel, RoyBoy, Matteh, Thunderbrand, TMC1982, Devil Master, Deathawk, Func, Diceman, Audrey, Anonymous Cow, FredOrAlive, Sukiari, Jason One, A strolling player, Bob rulz, CyberSkull, Jtalledo, Demi, MarkGallagher, InShaneee, Velella, Helixblue, Peter McGinley, Dremora, Alai, Red dwarf, Mahanga, Marasmusine, Veemonkamiya, Alvis, Jeffrey O. Gustafson, Woohookitty, Brazil4Linux, TigerShark, Percy Snoodle, Millard73, Jeff3000, Clemmy, Hbdragon88, ThomasHarte, Combination, Marvelvsdc, Sneakums, Mandarax, Xizer, Electricmoose, David Levy, RadioActive~enwiki, DePiep, Grammarbot, Rjwilmsi, Aelveric~enwiki, Nick R, Petree, Sango123, A Man In Black, Dionyseus, FlaBot, Moskvax, Ian Pitchford, SchuminWeb, Chanting Fox, Gurch, Czar, Jonny2x4, Okto8, UnlimitedAccess, Joerger, Chobot, Dstln, Korg, Igordebraga, Quicksilvre, Jason.cinema, YurikBot, RobotE, Hairy Dude, Dannycas, Hyad, WAvegetarian, Anonymous editor, Markpeak, Chensiyuan, Tirian, Gaius Cornelius, Bhavinshah, Anomie, RattleMan, Buuneko, Snkcube, Smartyhall, Diotti, Larsinio, Nanten, Y2kevbug11, Bota47, N. Harmonik, ^o^CORVUS^o^, Thelaughingman, Ms2ger, Nin10dude, Nikkimaria, Chopper Dave, Closedmouth, Alakazam, Th1rt3en, Jecowa, BlazeHedgehog, DKH, JLaTondre, Silvergunner, NFG, DCEvoCE, Doom127, Manmonk, Ryūkotsusei, That Guy, From That Show!, SmackBot, Khfan93, F, Renegadeviking, Translucid2k4, DuoDeathscyther 02, Mr. Pointy, Germ~enwiki, Unyoyega, C.Fred, Cutter, Jagged 85, Thunderboltz, Alan McBeth, Darklock, Ian Rose, MPD01605, Slo-mo, Chris the speller, GoldDragon, 32X, Onesimos, Lone Guardian, Kfroog, Moshe Constantine Hassan Al-Silverburg, RexImperium, Mark7-2, KieferSkunk, Mawich, Alphathon, OrphanBot, Sephiroth BCR, Rrburke, Nyletak, Super box A, TheAxeGrinder, Davelukeford, Swaaye, Filpaul, FreeMorpheme, Curly Turkey, Farm Zombie, Mr Stephen, Va.va, AxG, Ryulong, TPIRFanSteve, EEPROM Eagle, Saxbryn, Koweja, Animedude360, Phuzion, Calysma, TJ Spyke, SubSeven, TwistedArachnid, Mvent2, Hadoken, Blakegripling ph, IvanLanin, IvanDíaz, Arwengoenitz, Keelhaul, Az1568, Hadoken2000, RdCrestdBreegull, Offensiveandconfusing, CmdrObot, Plainnym, Cyrus XIII, Benjaminjoel322, Nczempin, Mika1h, ZachReed, Jesse Viviano, Juhachi, Daddy luv, Skybon, Auger Martel, Fruitbatnt, Djsonik, Cydebot, Bemo56, Chsea234, Gogo Dodo, GMTV, Soetermans, Dekabreak, Dancter, Rhe br, Thaddius, UberMan5000, Guyinblack25, Col. Hauler, Thijs!bot, Lord Hawk, N5iln, NsdrNPC, Marek69, X201, Dannyjonejules, Silver Edge, Caleson, AntiVandalBot, Luna Santin, Bull-Doser, Scepia, RobJ1981, TTN, Dragon DASH, Jhsounds, J'onn J'onzz, BrutalPootle, CPitt76, PresN, Dreaded Walrus, Fennessy, Tengu99, HellDragon, Hut 8.5, TAnthony, Jigahurtz, Y2kcrazyjoker4, .anacondabot, Hasek is the best, Gamkiller, Janadore, Ecksemmess, Avicennasis, Gnu andrew, Robotman1974, Nreive, Thibbs, Robivy64, GRAND OUTCAST, Inclusivedisjunction, B. Wolterding, An Sealgair, Gwern, Cube b3, Xtreme racer, Neorococco, Juansidious, R'n'B, CommonsDelinker, Yaca2671~enwiki, Nevakee11, Richiekim, Davebrck, Xenoranger, LedRush, 123wiki123, Kulmala, Paranoia Agent, Dispenser, Radicalfaith360, BtEtta, Awesomej1000, Guns2006, Guru Larry, S, Mr Wesker, Izno, Buggedoutnicy, CardinalDan, RingtailedFox, Mike Yaloski, Sjones23, Sarenne, Sanshiro, IllaZilla, Iobus55, Barroids, Jungle King, Ecopetition, Gravey9, Mick aka, Brianga, Helaynehag, Dirtyharrydi, Maverickhunterz, Hattes, Ipwnthecrusades, Theaveng, Oxymoron83, BenoniBot~enwiki, Fratrep, Macy, JohnnyMrNinja, Wonchop, Veraliton~enwiki, ImageRemovalBot, Fecman94, Martarius, De728631, ClueBot, Snigbrook, Malpass93, Badger Drink, EoGuy, Corbie33, AR Argon, ElectricalTill, Red Phoenix, Aylesburyape, Scatman2007, TBustah, MyMii, ChaosAngelZero, Bokan, Cawunited, 718 Bot, JCC87, Alexbot, DanielPharos, Thingg, Project FMF, Dank, Tezero, Classicrockfan42, DumZiBoT, InternetMeme, XLinkBot, Spitfire, Gallodannyo, Feinoha, Ost316, WikHead, Maxxfarras, Nuxius, Gazimoff, SelfQ, Addbot, RandySavageFTW, Qqkachoo, Megata Sanshiro, RedRose333, GD 6041, Darkness2005, CactusWriter, Cst17, Download, Robotriot, Leucius, DreamHaze, FrysUniverse, Tassedethe, Lightbot, Teles, Shinobi MVP, Zorrobot, Count druckula, Legobot, Luckas-bot, TheSuave, Yobot, Crizzlec25, Narodniheroj, Washburnmav, SegaSaturnUK, Jonathon43, USAJAP1, AnomieBOT, Calebbozeman, Message From Xenu, Laserdisc, Jim1138, Mcjakeqcool, Danno uk, Citation bot, LilHelpa, JElielx, FreeRangeFrog, HDS-GTR, Xqbot, NeoDoubleGames, Warmpuppy2, Ubcule, 7om, Sergecross73, Gbruin, Gameconsoleguide, FrescoBot, Tug97, Glider87, NGSF, Kwiki, Purpleturple, W 2465a67df, Lucia Black, Aizuku, Ryo Suzuki, Secret Saturdays, Picture-

House harryland, DellTG5, Lightlowemon, Black Squirrel 2, Mariacer Cervantes, Martin IIIa, Nogib, Fayedizard, PleaseStand, Reach Out to the Truth, RjwilmsiBot, DexDor, EmausBot, John of Reading, WikitanvirBot, AmericanLeMans, G&CP, SexyKick, Erpert, Oussama2002, Meicyn, Hydao, Neh0000, H3llBot, Railer-man, MAINEiac4434, Thekeyboardman, BookDen, Asdfsfs, JHoningh, ChuispastonBot, ThePowerofX, GermanJoe, Evan-Amos, Vittupaska123, Onpon4, Chrisfjordson, TheTimesAreAChanging, ClueBot NG, KazamaBoy, Cambino, NGMan62, Satellizer, DarkStar1997, Egg Centric, Juan C. S. H., BluRayStation3, Capsoul, Helpful Pixie Bot, Waterloosunset27, Lowercase sigmabot, BG19bot, TheLoverofLove, Winxclub93, Duhy132, AramilGaia, MusikAnimal, NukeofEarl, PhilipTerryGraham, Zeke, the Mad Horrorist, MonkeyKingBar, Harizotoh9, Thebigs14, The1337gamer, BattyBot, It's Numeric, ChrisGualtieri, Arcandam, Jimmy the rustler, StevefromQuebec, Dissident93, SoledadKabocha, ABunnell, Jeppestrolleri, BDE1982, Paspie, Backpacks and Shopping Bags, Echennessy, Vanished user lalsdi45ijnefi4, SegaKing247, Recapatcha, Master56456, Batamamma, Kevinfrombk, Gixce93, Jbuc14, John7Tree, Oranjelo100, VanishedUser sdu9asdsopas, Ethereal Static, Monkbot, Zer0Phi, Coreyray1000, KombatPolice, Retroking1981, Chan-Murphy, TheKingsTable, BethNaught, John Mayor ERS, Master Powerade, BustaBunny, TheGeneralofWar, TheRealAfroMan, MarioSonicU, Sephirothkefka, GodOFGamers72, Dr. Pieman, FACBot, Cartakes, Psycho santa, Jon the VGN3rd, BD2412bot, Maiathebest7878787878, Cubisticmage and Anonymous: 695

- **Sega** *Source:* https://en.wikipedia.org/wiki/Sega?oldid=680670057 *Contributors:* Paul Drye, Eloquence, Jzcool, Malcolm Farmer, Aldie, SimonP, Atlan, Frecklefoot, The T, Wapcaplet, Tregoweth, KAMiKAZOW, TUF-KAT, Angela, Darkwind, Lupinoid, Nikai, Ghewgill, Andrevan, Tedius Zanarukando, WhisperToMe, Wik, Roadmr, Tpbradbury, Furrykef, K1Bond007, Dan Mazurowski, Tonius, Omegatron, Shizhao, RadicalBender, Phil Boswell, Robbot, Fredrik, Yelyos, Hadal, Bbx, GreatWhiteNortherner, DraQue Star, JamesMLane, DocWatson42, Jacoplane, 0x0077BE, Nifboy, MMBKG, Misterkillboy, Ferkelparade, NeoJustin, Chrishill61, FriedMilk, Xinoph, Fourlittlediamonds, Bobblewik, Wiki Wikardo, Golbez, SonicAD, Paraiba, Tom k&e, Chowbok, Utcursch, Kusunose, Mamizou, Hi, Jeff. Hi!, Rdsmith4, D3v4st4t0r, Bodnotbod, Halo, Sam Hocevar, Neutrality, Jh51681, Trevor MacInnis, Zoganes, Lacrimosus, RevRagnarok, Heegoop, Discospinster, Rich Farmbrough, Avriette, Vague Rant, Hydrox, Qutezuce, Clawed, Pixel8, Waka, Prion~enwiki, Pavel Vozenilek, Shadow Hog, Indrian, Bender235, ESkog, Evice, Ruyn, Shanes, TMC1982, Devil Master, Bobo192, Stesmo, NetBot, ERW1980, Jporter07, Sabretooth, Giraffedata, Jerryseinfeld, Sasquatch, Jason One, WideArc, Arthena, Geo Swan, Jtalledo, Andrewpmk, Riana, Fritz Saalfeld, Goldom, Celzrro, Dalm, Snowolf, TheRealFennShysa, Javacava, FA010S, Zxcvbnm, CherryMay, Alai, Drbreznjev, Kitch, KUsam, MickWest, Angr, Firsfron, Alvis, Woohookitty, Jackel, PoccilScript, Ritz, Bratsche, Robert K S, Gapporin, NeoChaosX, Hbdragon88, Awk~enwiki, Crazysunshine, TheEvilBlueberryCouncil, ThomasHarte, Daniel Lawrence, Banpei~enwiki, Combination, Alrik Fassbauer, Dysepsion, MassGalactusUniversum, Graham87, Xizer, BD2412, David Levy, JIP, RadioActive~enwiki, Rjwilmsi, Tim!, Nightscream, Demian12358, Perks, Eugeneiiim, B'man, Browned, Aelveric~enwiki, JoshuacUK, Ukdan999, Nick R, Vegaswikian, CQJ, AceTracer, Oo64eva, A Man In Black, FlaBot, King Dedede, SchuminWeb, Weebot, CR85747, MattFisher, Pumeleon, Krackpipe, Ayla, Mitsukai, Valermos, Czar, Stormwatch, D.brodale, Cdbarker, ...adam..., Chobot, Dstln, Rikoshi, Nitoplayer, Igordebraga, Bgwhite, ShadowHntr, Kakurady, YurikBot, Wavelength, Sceptre, I need a name, NTBot~enwiki, Beltz, Dannycas, RussBot, Petiatil, Mee Ronn, TheDoober, Sasuke Sarutobi, IanManka, Subsurd, Stephenb, Gaius Cornelius, CambridgeBayWeather, Wgungfu, Rsrikanth05, Wimt, Rhindle The Red, Royalbroil, Knyght27, NawlinWiki, Anomie, WulfTheSaxon, Smash, Pagrashtak, RattleMan, NW036, Tfine80, SirNuke, NP Chilla, Confero, Irishguy, Saoshyant, Brandon, Inhighspeed, DAJF, RFBailey, PhilipO, BattleMario, Kaiti, Nanten, FlyingPenguins, Palpalpalpal, Gamelore, Goodcow, Golladayp, Antoshi, Rwalker, Starze, Denis C., JaimeyWB, TimK MSI, Groink, Zelikazi, Xino, Alpha 4615, TransUtopian, J. Nguyen, Wpollard, 21655, Nezuji, Barryob, Nikkimaria, Chase me ladies, I'm the Cavalry, Closedmouth, Willirennen, Nolanus, Luckybolt, Shawnc, PureLegend, Danikat, Kevin, Lando242, DisambigBot, ViperSnake151, Katieh5584, Kungfuadam, Ief, JosephLondon, Kazmeyer, Airconswitch, Jeff Silvers, NiTenIchiRyu, Stumps, DVD R W, Luk, SmackBot, Antster1983, Lashiec, Khfan93, Moeron, Nihonjoe, Renegadeviking, DuoDeathscyther 02, Liddo~enwiki, Herostratus, MarjorieCook, Borincano75, Rjsec4ever, Pgk, Jagged 85, Stifle, Knilt, Delldot, Rojomoke, Jipcy, Doc Strange, Geoff B, SonicLifeform, Edgar181, Gilliam, Sbonsib, Seann, DividedByNegativeZero, Doktor Wilhelm, Cabe6403, Parrothead1983, Chris the speller, Bluebot, Keegan, Quinsareth, Landrjm, MK8, 32X, Jnelson09, Master of Puppets, Thumperward, Onesimos, Kfroog, MalafayaBot, Jerome Charles Potts, FordPrefect42, Baa, CMacMillan, Raymie, Colonies Chris, Nintendude, MaxSem, Can't sleep, clown will eat me, MisterHand, Killerclaw, Alphathon, Viperphantom, Thebeast666, Efitu, OrphanBot, Folksong, Azumanga1, Greenshed, Icrofoot, Yosha, Logan GBA, PrometheusX303, Super box A, SammyGreen, Crv1, Tehw1k1, S@bre, Baloonda, Salamurai, Vina-iwbot~enwiki, Yuri Elite, WayKurat, Ohconfucius, Guyjohnston, Rick Browser, Mchart, Cjcamilla, Palillont, AllStarZ, Lazylaces, Breno, Stefan2, IronGargoyle, Alpha Omicron, EnthusiastFRANCE, Enelson, Squeak90, 16@r, MarkSutton, Feureau, Incognit000, RememberMe?, Cyanidesandwich, Atirage, Ryulong, Renhazuki, AEMoreira042281, TJ Spyke, SubSeven, MikeWazowski, Chris Price, Otduff, Iridescent, Alimn, Wfgiuliano, JoeBot, Ashura96, IvanLanin, DavidHOzAu, Amakuru, Keelhaul, Esurnir, Majora4, The Music of Magic, Az1568, FairuseBot, Tawkerbot2, DKqwerty, Sjm757, SkyWalker, J Milburn, CmdrObot, Tanthalas39, Le poulet noir, Lavateraguy, Cyrus XIII, Addict 2006, Nczempin, Mika1h, Dbzsamuele, Superspam111, Kev19, RockMaster, Erencexor, Auger Martel, Foxcat, Markallangibson, Slazenger, Cydebot, Raamin, Farine, Cambrant, Gunstarhero, SyntaxError55, Gogo Dodo, Pig de Wig, Hebrides, Corpx, Adolphus79, Dancter, Neoforma, DumbBOT, Viking uk, UnDeRsCoRe, Optimist on the run, Bungle, Kozuch, Zalgo, RedWolfX, Aldis90, Thijs!bot, Lord Hawk, Seasponges, GentlemanGhost, Kablammo, Voracious reader, Mojo Hand, Sushi-x, John254, X201, Raveeshworldwide, JustAGal, Mnemeson, Sir Simon Tolhurst, Lowercase, Thljcl, Grayshi, Silver Edge, Scottandrewhutchins, Trace The Hedgehog, Mentifisto, AntiVandalBot, G1m2, Elven6, Seaphoto, Drewdy, Prolog, Dragon DASH, Jhsounds, Vendettax, Svette2002, Spartaz, Jimeree, Arx Fortis, Kaini, PresN, Radio Flyer Guy, Student Driver, Kcowolf, Jt 200075, JAnDbot, Husond, Janus657, ThomasO1989, Zebbe, Jason Stormchild, Andonic, Hut 8.5, East718, Dream Focus, MSBOT, Kirrages, D3athstardisco, Kerotan, Madomen, Geniac, Kakarott, Magioladitis, Bongwarrior, VoABot II, Neofcon, TheAllSeeingEye, Kuyabribri, JNW, Janadore, Tsukento, Galifardeu, ShadowTao, Ibrahim s, Dovereg, Gr1st, Rob Enduro, Fallschirmjäger, Animum, VegKilla, Cliché Online, Allstarecho, Mareimbri, Edward321, ChazBeckett, Nintenboy01, ElKameleon, Brittany Ka, FisherQueen, Blacksqr, Cube b3, Ultrahead, SpecialWindler, MartinBot, Grandia01, Letni, Tgeairn, Erkan Yilmaz, BGOATDoughnut, Aureez, J.delanoy, Kimse, Bongomatic, Calamity-Ace, Ali, Aliquidparadigm, Fulou, WarthogDemon, Digital Man, Kenshinflyer, Sega31098, Bobisgreat, McSly, Duhman0009, Thomas Larsen, NiGHTS into Dreams..., Plasticup, NewEnglandYankee, Sheardogers, SJP, Frogacuda, Johm000, Cmichael, Shadow Android, White 720, KylieTastic, Cometstyles, Guru Larry, Useight, Haddyrikabi34, Mr Wesker, KGV, Tkgd2007, Idioma-bot, Funandtrvl, Mwlin1, Vranak, SEGADREAMCAST, TreasuryTag, Thedjatclubrock, Jamcib~enwiki, Jeff G., AlnoktaBOT, Jdchamp31, Asrabkin, Sjones23, Tomer T, HelmsC1978, Philip Trueman, TXiKiBoT, Yotyu, AxellSlade, Radiuz, Mr. Top Hat Magic, Zacarrell, Adamodell, Mario 5757, A4bot, Hqb, SeanMooney, Inthenameofgod, GcSwRhIc, G4rce, Irtehax, Bmg916, Lradrama, IllaZilla, AtaruMoroboshi, Camel Light Flighters, Hashmasta, George's Glowball, Grease Guy, Billinghurst, Sonicobbsessed, Rwell3471, Mick aka, Sesshomaru, The Devil's Advocate, DSFanatic, Twooars, Asim18, Teddy.Coughlin, AlleborgoBot, Munci, Fdiddy, Enc Company Agent, ConnTorrodon, Kastrel, Overlord11001001, ZOMGitzSEAN, Bigbadjon101, Tiddly Tom, Hattes, Sparrowman980, Panenforcer, VVVBot, Phanink, Ddddbbbb, The very model of a minor general, Edgewrth, S200048, Mr.Z-bot, BlueAzure, Theaveng, Oda Mari, Segata128, King-

GreyWolf, Oxymoron83, UltraNintendoEntertainmentSystem, Nuttycoconut, KPH2293, Spock2266, AnonGuy, Lightmouse, R. C. Mongler, Manway, BenoniBot~enwiki, Fratrep, Gunmetal Angel, Gamefan inform, OKBot, Info845, Anchor Link Bot, TaerkastUA, Johnnywilbur, Wonchop, Kn00tcn, Kanonkas, Rogue Commander, Fruit Punch Star, Mega Tele-Funk's Hi-Fi 5000, ImageRemovalBot, Videogamessuck, Spiderverse, Mr. Granger, Martarius, Sfan00 IMG, ClueBot, Together Forever Guy, GorillaWarfare, PipepBot, The Thing That Should Not Be, Madskunk, Raffage, Mx3, Parkjunwung, Red Phoenix, Rhonin the wizard, Drmies, THEGREATMADMAN, CounterVandalismBot, Slayer Guy, Magiciandude, Whiteguysamurai, StigBot, Harland1, The luigi kart assasions, Auntof6, Smackboy69, Unreal221, Sharp point, Jimmyc85, Jonnyclavin1616, PatLTornado, Excirial, Mynameisnotpj, Erebus Morgaine, TheCharlyHorse, Resoru, Adimovk5, Leonard^Bloom, Kaiserjagen, Flippers~com, NuclearWarfare, Mcryan0, DanLozanovski, Thehelpfulone, Another Believer, Gyozilla, Thingg, Project FMF, Aitias, 2, Tezero, Kokoro20, Classicrockfan42, DumZiBOT, CBMIBM, XLinkBot, Terriblefish, Flywick7, Spitfire, Jurox51887, Mrmusic16, Luckysidgem, Rror, Gallodannyo, Dthomsen8, Ost316, PeruAlonso, Little Mountain 5, NellieBly, HarlandQPitt, FightingStreet, Gameplaya007, Bobby 67897474758, Anticipation of a New Lover's Arrival, The, Irysa, MatthewVanitas, Addbot, Tcncv, SSB Fan, GM25LIVE, Fielddaysunday, Caramon94, Cartin91, CanadianLinuxUser, Mac Dreamstate, Crris08, LaaknorBot, ShepBot, Glane23, Leucius, Nickin, AndersBot, AnnaFrance, DreamHaze, Favonian, 5 albert square, ChartreuseCat, Tyw7, Moothat2, Sonicthehedgehog9000~enwiki, Krano, MiltonP Ottawa, Shinobi MVP, Theprophet08, MuZemike, Legobot, Yobot, Unknown the Hedgehog, Bocafan76, Themfromspace, Ptbotgourou, Legobot II, II MusLiM HyBRiD II, I didn't push her, Terrifictriffid, Giusex27sc, Nallimbot, AnomieBOT, DemocraticLuntz, Retardnationfan, Caprinoe, Ibssum dal abdur, Piano non troppo, TParis, Germanname1990, Mtasf, Bonusballs, Materialscientist, NINJA BLADE, RobertEves92, Maniadis, GB fan, ArthurBot, LilHelpa, Gsmgm, Daftpunkboy93, GotFilk, Xqbot, APDzie, NeoDoubleGames, Sionus, Capricorn42, Kody-the-Fox, Acebulf, Junkcops, MatheusBond, Pelerin2, Jankuza, Armbrust, NickelKnowledge, Sergecross73, Brandon5485, RibotBOT, Dark ressurection, Aperson1234567, F.Pavkovic, Maxynator123, Shadowjams, Noname1234321, Sesu Prime, Jerrysmp, Thehelpfulbot, Xboxgeek, FrescoBot, SharpTurn64, Zzoidberg, Ndboy, 1779Days, Yosoy2kool4u, Tug97, HJ Mitchell, Agbwiki, Boleyn3, Kwiki, Purpleturple, DivineAlpha, Jackmelody95, Andrew wu77, Kuover, FriscoKnight, DrilBot, Abani79, Aizuku, Tanweer Morshed, Jonesey95, Tinton5, Ftc08, Spongesonic277, Thinking of England, DBallow, Ryo Suzuki, Secret Saturdays, Salvidrim!, Heavydata, Nerefis, Lightlowemon, FoxBot, Mundilfari, Memory Prime, Martin IIIa, Lotje, Carniolus, Phil A. Fry, Gthec9909, Diannaa, Superblackjack, Reach Out to the Truth, Bobby122, Adamcly, Mean as custard, The Utahraptor, Ontarget777, The Stick Man, Petermcelwee, EmausBot, John of Reading, Bykfiend42, Dewritech, RToren, GoingBatty, RA0808, G&CP, Emo-tional being, NotAnonymous0, Somebody500, Mz7, Alexandrus2010, Illegitimate Barrister, Hofstader, SEGA Casino, Bartman01234, Bighouseinthecountry, ElationAviation, ExtremeGiga, Lacon432, Tomic The Hedgehog, Bilbo571, H3llBot, ArdiPras95, Wani, MR.Nintendo13, DavianThule, SporkBot, Galaxy0518t, Dumitru Gherea, SkyBon, Thekeyboardman, MOMBO BAGGINS, L Kensington, Lord Psyko Jo, Donner60, ChuispastonBot, Evan-Amos, Jse11, Kai445, TheTimesAreAChanging, Ao25798, Faramir1138, ClueBot NG, T3hyoshi, Prioryman, Horseman16, Santtu37, Awesomeness95, VanishedUser sdu8asdasd, Despatche, O.Koslowski, ScottSteiner, Widr, Chris65536, Vidpro23, Helpful Pixie Bot, IPawdlol, Terminated Joker, Calidum, Pochop53, Superglove, PeladonFeo, Cjkaminski, 20chances, Lowercase sigmabot, BG19bot, Citizenjoe100, M0rphzone, WikiKong, Cliffeed, Dancarblog, Pdiddyjr, Chubzhac, Compfreak7, Ushio01, Jellotinerage, ACBContent, September 1967, BigAussieMonster, Pasq243, Ultraultragaben, Sonichuchris, Sanic133, Pomfersgonnapomf, DPL bot, Stoopdapoop, MartinZwirlein, Glacialfox, Segamegadrive32x, JC DENTON SMOKES WEED, Austrollolol, The1337gamer, Downunder112, Mahboi1, Hansen Sebastian, Somebody101, I eM NICK, Fluffystar, Pratyya Ghosh, Themariobrosadvance, Black Agent, That's Bupkis, Dobie80, MadGuy7023, Malcolm L. Mitchell, Manbuff12345, Dissident93, Taffysaur, Codename Lisa, Mogism, Idcwhatusernameis, Guiletheme, In87HueyReleasedThis, 81M, Cooper122, Elclappo, Elclapo, Lugia2453, Frosty, SFK2, KahnJohn27, Awsom1324567, Wiki Invader97, Jusmoore123, PC-XT, Faizan, Amier42, Seqqis, Youmeanittasteslike, NikitaBoy, ShadowInuYusei, Vanished user lalsdi45ijnefi4, SegaKing247, VanishedUser sdu9aya9fs232, Lemaroto, Hoppeduppeanut, Kevinfrombk, Ycooldog, HannahPavina, Asmetr, Кирилл Ерин, Cheeseisdisgusting, Dogcat426, Uncle ben, Kind Tennis Fan, Hisashiyarouin, Silverstreak Folf, Ganthony4186, HelloWorldCanIPeeOnYou, Lemonater47, SS7 Somebody, Imesswithpagesguy, Hades173, Thephil12312, OneSidedViewpoint, 1234567890w0987654321, SHITgun user, Monkbot, Thibaut120094, Ilovehedgiessonicnusa, Noa dave, StephenCezar15, AustralianPope, ClutchStatic, Folinator5000, Captainsanfrancisco, Lor, Amortias, DoveChaser, AdrianGamer, ClassicOnAStick, Aidan68945, Sonic N800, MarkToader 965, Tripple-ddd, Ronnie the speedster, Lunamightmoon, Historyman2112, Wariostarx, Thekev779, YeOldeGentleman, Thatoneguy1010, ToonLucas22, CAS222222221, Roman Isolated, LotspagHero, JasonBSinger, BlusterBlaster, Electricninja2234, Wikipicky1090, Alessiio155ti, TomD200, HahaLOL696969, Oriam421, Swaggosaurusrex47, ProprioMe OW, CAPTAIN RAJU, Spiral knights is the best game ever, Attacker121, Soothaxial, Zeroshift3000, SethNr.78 and Anonymous: 1501

- **CD-ROM** *Source:* https://en.wikipedia.org/wiki/CD-ROM?oldid=678869503 *Contributors:* Damian Yerrick, Magnus Manske, Dreamyshade, Tarquin, Koyaanis Qatsi, Andre Engels, Rgamble, Nate Silva, Ghakko, William Avery, Ben-Zin~enwiki, Drbug, Heron, Frecklefoot, Patrick, Tim Starling, Willsmith, Modster, Dante Alighieri, Mahjongg, Nixdorf, Pnm, Liftarn, Gabbe, Wapcaplet, Ixfd64, Geoffrey~enwiki, Ahoerstemeier, CatherineMunro, Basswulf, Glenn, Whkoh, Poor Yorick, Nikai, Kwekubo, Emperorbma, Crissov, Dcoetzee, Reddi, Saltine, Nv8200pa, Tempshill, Omegatron, Wernher, Shizhao, Vaceituno, Flockmeal, EikeF, Robbot, Chealer, Bmaisonnier~enwiki, Hadal, Guy Peters, Hvs, Matthew Stannard, Giftlite, DocWatson42, Kim Bruning, Kenny sh, Ferkelparade, Mcapdevila, Kccricket, Jason Quinn, Macrakis, SWAdair, Bobblewik, TheMaestro, Barneyboo, Andycjp, CryptoDerk, Mako098765, Quarl, Bumm13, Kevin B12, Sfoskett, Sam Hocevar, Sillydragon, Aidan W, Slady, Econrad, Discospinster, Rich Farmbrough, Rhobite, FloSch, AlanBarrett, Pavel Vozenilek, Hhielscher, Evice, CanisRufus, El C, Kwamikagami, Dennis Brown, Triona, Fourpointsix, Bobo192, Mr2001, Wikinaut, Idleguy, Nsaa, Espoo, Alansohn, PatrickFisher, Sade, Water Bottle, Lightdarkness, Marianocecowski, Angelic Wraith, Luspari, Dtcdthingy, Ceyockey, Kbolino, Bruce89, Zntrip, Ondrejk, Chardish, Anilocra, Rocastelo, Ruud Koot, Dah31, Sega381, Byronknoll, Banpei~enwiki, Reisio, Rjwilmsi, MJSkia1, Nightscream, IRT.BMT.IND, Panoptical, Vary, Amire80, Linuxbeak, Lordkinbote, SMC, Oblivious, Bhadani, Husky, StuartBrady, Mariocki, Gurch, Jump3R, Intgr, Synchrite, Pip2andahalf, Phantomsteve, FrenchIsAwesome, Sillybilly, Phantombantam, Stephenb, Gaius Cornelius, Pseudomonas, Bovineone, Wimt, Stassats, NawlinWiki, Emuroms, Grafen, Irishguy, Hugh Bennett, Matticus78, ScottyWZ, Zwobot, Jeh, Izcool, Sebleblanc, CLW, Yudiweb, Navstar, TheSeer, Sperril, Closedmouth, GraemeL, Fourohfour, Poculum, GrinBot~enwiki, Nick-D, Fluxpattern, SmackBot, Estoy Aquí, KocjoBot~enwiki, Jagged 85, Thunderboltz, Pennywisdom2099, BiT, Cached, Gilliam, Dell boy, Skizzik, Cs-wolves, Chris the speller, Thumperward, Oli Filth, Pavlina2.0, Jerome Charles Potts, ACupOfCoffee, Rlevse, Jnavas, FiftyNine, Can't sleep, clown will eat me, Frap, OrphanBot, OSborn, Born Acorn, Rrburke, Kcordina, Achilles2.0, Safwankenobi, Nakon, Acdx, LeoNomis, TenPoundHammer, Downsjn, Kuru, John, J. Finkelstein, Flip619, Edwy, RomanSpa, PseudoSudo, Craigblock, Loadmaster, SQGibbon, Feureau, Dicklyon, Mouseboy, Zachdms, Waggers, Chaos Reaver, Spiel496, Avant Guard, BranStark, Bartekfm, Dp462090, Aeons, Tawkerbot2, Ouishoebean, Emote, Zarex, Woudloper, Ninetyone, JohnCD, Aussiepete, Sorn67, Mhs5392, JVinocur, Necessary Evil, Cydebot, Gremagor, Meno25, Wpillar, Hebrides, Luckyherb, Asenine, Dinominant, Omicronpersei8, Thijs!bot, Kubanczyk, Rrose Selavy, N5iln, Jastein, Marek69, Davidhorman, Escarbot, Rees11, AntiVandalBot,

Luna Santin, Seaphoto, Turlo Lomon, Edokter, Smartse, Beowulf6561, Res2216firestar, Eng101, JAnDbot, Deflective, Yurimxpxman, Barek, Dream Focus, Karlhahn, Bongwarrior, VoABot II, Swpb, Tedickey, Shablog, Aka042, BrianGV, Slowacki, Comstad, 28421u2232nfenfcenc, Hdt83, MartinBot, Mpwheatley, Roastytoast, Eloopj, Ceros, J.delanoy, Public Menace, StuThomas, Dispenser, Mahewa, SJP, Juliancolton, Cast123, VolkovBot, Cireshoe, ABF, Mrh30, Robtanner, Epson291, Philip Trueman, DanMatthewsUK, Benjamin Barenblat, TXiKiBoT, Arnon Chaffin, Arydberg, Seb26, ITurtle, Mannafredo, This acccount is 4 vandalism, Haseo9999, VanBuren, Brianga, Mary quite contrary, Pjoef, Thunderbird2, Logan, Demize, EmxBot, D. Recorder, Demmy, Mehmet Karatay, Tharcore, The Random Editor, SieBot, Caulde, Nestea Zen, Caltas, Matthew Yeager, Keilana, Bentogoa, Flyer22, Theaveng, Lightmouse, Fratrep, Escape Orbit, Kanonkas, ClueBot, C xong, Avenged Eightfold, The Thing That Should Not Be, Starkiller88, Jdgilbey, Manogod0605, CounterVandalismBot, September 11 terrorist, Blanchardb, Ridge Runner, Excirial, Muenda, Rhododendrites, Skytreader, Thingg, Versus22, Freedomtroll, Pichpich, Rror, Ost316, Skarebo, NellieBly, Noctibus, 死亡, Lanky217, Addbot, Ghettoblaster, AVand, Hda3ku, M.nelson, Ronhjones, LaaknorBot, Favonian, Brenden mahon, Zorrobot, Jarble, Luckas-bot, Yobot, Playclever, ArivaldH, II MusLiM HyBRiD II, Crispmuncher, Nallimbot, SirWizard, LongAgedUser, AnomieBOT, HairyPerry, Ciphers, Jim1138, IRP, Piano non troppo, Kingpin13, Visva95, Maxis ftw, GB fan, Wikante, Jeffrey Mall, Hanberke, Timme-southpark, Julia-The-Little-Lady, Ched, GrouchoBot, Frankie0607, RibotBOT, Grozsa11, Shirtez13, Shadowjams, WaysToEscape, Krinkle, MISTYFAN4EVER8887, Fcjefe, I3elphegor, Ctech72, Elockid, Arlo.Clauser, Maddaye, Smuckola, RedBot, Pikiwyn, SpaceFlight89, Killergod129, Sillik, Marcopete87, Vkil, Lotje, Weedwhacker128, Some Wiki Editor, DARTH SIDIOUS 2, Mindy Dirt, The Stick Man, Salvio giuliano, EmausBot, John of Reading, Edaddy13, Dewritech, Racerx11, GoingBatty, Compugeekmsn, RA0808, Majed, K6ka, Heritage.john, H3llBot, Unreal7, JoeSperrazza, Jay-Sebastos, Bomazi, Evan-Amos, Minority Carrier, Socialservice, Wikepidia is wrong, Georgy90, ClueBot NG, Msunderland, Matthiaspaul, Eggman2011, VIPclubmaster, Doh5678, WikiPuppies, Tbz1997, NUMB3RN7NE, PhnomPencil, Wiki13, Michaelmalak, Trondeg74, Paweł Ziemian, PatheticCopyEditor, Theone666, EagerToddler39, Mogism, Lugia2453, Vivruth, Bananasoldier, Mphenley, EntertainmentAssociates, Ginsuloft, ‏גלדון‎, Monkbot, LePerfectionniste, Jorgecg13, BlueFenixReborn, Ffdddsf, KasparBot and Anonymous: 601

- **SuperH** *Source:* https://en.wikipedia.org/wiki/SuperH?oldid=671997611 *Contributors:* AlexWasFirst, JakeVortex, Stan Shebs, Julesd, Sanxiyn, Kaal, Chealer, Alain~enwiki, David Gerard, Takanoha, Kenny sh, Solipsist, Madoka, Mamizou, Master Of Ninja, Now3d, Damieng, Moxfyre, Cdworetzky, Foobaz, JeR, Diceman, Trevj, Guy Harris, RuiPaulo, Conan, Bookandcoffee, Alecv, MarkusHagenlocher, ThomasHarte, Toussaint, FlaBot, Nihiltres, Intgr, Chobot, Bgwhite, Mordamir, ShadowHntr, Pelago, Thanurmi, Foofy, SmackBot, McGeddon, Sloman, 32X, Frap, Virtualsim, WhosAsking, Dicklyon, EdC~enwiki, Saxbryn, TJ Spyke, Alexthe5th, W heer, Jesse Viviano, MarsRover, Mschlett, Rhe br, Omicronpersei8, Thijs!bot, RCHM, Dulciana, Wmat, Peter Write, R'n'B, Yaca2671~enwiki, Alastaird, Rdautel, Una Smith, Thunderbird2, Noveltyghost, Dojcubic, GMPX, Teh.cmn, ImageRemovalBot, Rilak, Eeekster, DanielPharos, Dsant, Addbot, Rkalm, Scientus, Alastairdent, Yobot, AnomieBOT, J04n, FrescoBot, Mrpgriffin, Traxs7, Jonpatterns, Rangoon11, Chester Markel, BG19bot, Comp.arch, Mman2112 and Anonymous: 75

- **Motorola 68000** *Source:* https://en.wikipedia.org/wiki/Motorola_68000?oldid=678067223 *Contributors:* Damian Yerrick, Tuxisuau, Brion VIBBER, Stephen Gilbert, Wayne Hardman, Karen Johnson, Maury Markowitz, Ellmist, B4hand, RTC, JakeVortex, Alfio, Egil, Stan Shebs, Kazuo Moriwaka, Nikai, GRAHAMUK, Arteitle, JidGom, Ww, Greenrd, Zoicon5, Furrykef, Rronline, Wernher, Stormie, Ldo, AlexPlank, Robbot, Murray Langton, Fredrik, Calimero, RedWolf, Modulatum, Zidane2k1, CajunLuke, Wikibot, Seano1, Rsduhamel, Takanoha, Centrx, Brouhaha, Lproven, DavidCary, Levin, Ferkelparade, Jonabbey, Jjamison, Patrickdavidson, Pascal666, AlistairMcMillan, Solipsist, Neilc, Jonathan Grynspan, Tsangc, DNewhall, MFNickster, Gaul, Bumm13, Marc Mongenet, Grunt, Rich Farmbrough, Pixel8, Xmachina, ArnoldReinhold, Bender235, CanisRufus, Bletch, Dgpop, Drhex, Cmdrjameson, Matt Britt, Colin Douglas Howell, DaveGorman, Estrus, Mote, Michael Drüing, Hohum, Klaser, Angelic Wraith, Schapel, ProhibitOnions, Wtshymanski, Postrach, Firsfron, Arneth, RHaworth, LOL, Ae-a, Ntg, Tabletop, Hailey C. Shannon, GregorB, ThomasHarte, Graham87, Qwertyus, Shadowhillway, Rjwilmsi, Krash, Dionyseus, FlaBot, Mirror Vax, LaurenceV, Kolbasz, Glenn L, Chobot, Bgwhite, YurikBot, TexasAndroid, Hairy Dude, Yuhong, Wgungfu, Lavenderbunny, Rat144, Thiseye, CecilWard, Speedevil, Richardcavell, Cedar101, Ed de Jonge, JLaTondre, GrinBot~enwiki, SmackBot, Aths, Chris Burrows, Jagged 85, Arny, Elrond 3097, Sloman, TimBentley, Computer Guru, Thumperward, KieferSkunk, Frap, Radagast83, Martijn Hoekstra, Anss123, Shirifan, RCX, Loadmaster, Jgrahn, KurtRaschke, Blakeyed~enwiki, Norm mit, Shoaib Meenai, TO11MTM, Jabjabs, Total Eclipse, Musashi1600, HenkeB, MeekMark, Arrenlex, Hga, Gremagor, ChristTrekker, Orion Blastar~enwiki, Rhe br, Bufdaemon, Pipatron, Thijs!bot, Lord Nightmare, Al Lemos, Electron9, X201, Rees11, Guy Macon, Uvaphdman, LegitimateAndEvenCompelling, Richiez, NapoliRoma, RastaKins, VoABot II, Kovan, Robivy64, Gwern, STBot, Gasheadsteve, R'n'B, CommonsDelinker, Irishclause, Adavidb, Acalamari, Amydet, Potatoswatter, Cmichael, Didier Misson, Scottwh, Jcea, Sarenne, Milan Keršláger, Andy Dingley, Sesshomaru, SieBot, Fnagaton, Miremare, Calabraxthis, Jp314159, Radon210, Theaveng, Lightmouse, Dojcubic, Svick, Dtvjho, CultureDrone, Anchor Link Bot, MarkMLl, ClueBot, Rilak, Mild Bill Hiccup, Niceguyedc, Shjacks45, Myztry, Sun Creator, Fairseeder, Pearce jj, Mac128, Fryinglizard, WikHead, NellieBly, Eleven even, Addbot, Magus732, MrOllie, Robert.Harker, Luckas-bot, Yobot, AnomieBOT, Dav.vire, Citation bot, Tripodian, Ubcule, Albert Non, Prari, FrescoBot, Liparius, Patronanejo, Tomekdcd, Hoo man, Ray G. Van De Walker, DARTH SIDIOUS 2, John of Reading, Sbmeirow, Bcaulf, Mulletsrokkify, Mikhail Ryazanov, ClueBot NG, Pantergraph, Matthiaspaul, Borgmcklorg, Delusion23, SeattleMurki, Lwink2, Helpful Pixie Bot, CitationCleanerBot, Cyberbot II, ChrisGualtieri, Briancarlton, Dissident93, Comp.arch, Birdman86, OMPIRE, RoundupResistance, TD712, Helios crucible, Orgelmeister and Anonymous: 214

- **Sega NetLink** *Source:* https://en.wikipedia.org/wiki/Sega_NetLink?oldid=674003390 *Contributors:* Booyabazooka, Kingturtle, K1Bond007, Boffy b, Rick Block, Bobblewik, Jonny, ScarredSun, Firsfron, ThomasHarte, Josh Parris, FlaBot, Lightsup55, DLoom, SmackBot, 32X, Karnb, Offensiveandconfusing, Mika1h, Evan1109, VoABot II, Mrbrazil, RingtailedFox, AlleborgoBot, Lightmouse, Red Phoenix, Mild Bill Hiccup, Felix the Hurricane, Sheeeeeeep, XLinkBot, ZadocPaet, Rylennel, Seganer, Junkcops, FrescoBot, Diannaa, Awesome McKillington, Rayman60, John of Reading, G&CP, Evan-Amos, ClueBot NG, Xranger60, Satellizer, Wartoyz, LOLTRAINS, MarioSonicU, Communal t and Anonymous: 24

- **Virtua Fighter 2** *Source:* https://en.wikipedia.org/wiki/Virtua_Fighter_2?oldid=680806551 *Contributors:* Conti, CohenTheBavarian, Indrian, Thunderbrand, Devil Master, Longhair, Angie Y., WTGDMan1986, InShaneee, Sagitario, Jjatria~enwiki, Dangerous-Boy, ADeveria, ThomasHarte, Combination, BD2412, Brianreading, Zotel, BradBeattie, Dstln, TnS, YurikBot, TW2k5, Anetode, Xino, N. Harmonik, Mike Selinker, RicardoC, Garion96, SmackBot, Seishiro Mixenex, Dessydes, Doktor Wilhelm, Parrothead1983, Chris the speller, NickSCFC, Sharpevil, Enker Dot EXE, Ravimakkar, Cuddy Wifter, 041744, Ryulong, Eliashc, TJ Spyke, Bentendo24, Mika1h, Cydebot, Conquistador2k6, Rhe br, BetacommandBot, X201, RobotG, Xeno, Y2kcrazyjoker4, Dekimasu, Janadore, Sniper 99, Cliff smith, Sp0, R'n'B, PocklingtonDan, Loekashe, Hamstermin, Mr Wesker, Varnent, WOSlinker, GDonato, AutocracyBot, JohnnyMrNinja, Hoobastank123, Nuxius, Addbot, Megata

Sanshiro, GD 6041, ‏مجيز فرتحم‎, Yobot, Erigu, Viking59, AnomieBOT, AarnKrry, RedBot, Martin IIIa, Lorson, HiW-Bot, Uniltìranyu, The-HeronGuard, Bill Hicks Jr., ChuispastonBot, TheTimesAreAChanging, Horseman16, Alexxxos, HMSSolent, C3F2k, DrRockso87, DC Malleus, Khazar2, SNAAAAKE!!, Homechallenge55, Ayşegül K. Yağız, Richolmes14, Kevieman94, Monkbot, Zer0Phi, MarioSonicU and Anonymous: 87

- **Sega Rally Championship** *Source:* https://en.wikipedia.org/wiki/Sega_Rally_Championship?oldid=680377891 *Contributors:* Shoaler, K1Bond007, Stewartadcock, Grm wnr, CALR, Loganberry, YUL89YYZ, StalwartUK, Sietse Snel, DaveGorman, Brainy J, Jamyskis, CherryMay, Kouban, Clemmy, Combination, BD2412, RadioActive~enwiki, Rjwilmsi, Nick R, The wub, FlaBot, Ultrasound, RussBot, Rambutaan, Mark Kim, Felsir, N. Harmonik, Shawnc, Rex Nebular, DuoDeathscyther 02, Jagged 85, Eskimbot, Commander Keane bot, Doktor Wilhelm, OrphanBot, Folksong, GVnayR, Ravimakkar, Teancum, EnthusiastFRANCE, Klango, Atirage, Ashura96, Mika1h, Zamamee, Cydebot, Rhe br, Alaibot, BetacommandBot, Thijs!bot, A is to B as B is to C, X201, PhonicsMonkey, Prolog, Fayenatic london, Dreaded Walrus, DOSGuy, .anacondabot, Janadore, Jklsemicolon, Redslap, R'n'B, Dispenser, Mr Wesker, Onore Baka Sama, Frees, Ponyo, Brenont, AutocracyBot, Martarius, Red Phoenix, Addbot, Dawynn, Megata Sanshiro, Lightbot, Megaman en m, Luckas-bot, Amirobot, USAJAP1, AnomieBOT, MrSaturn33, Junkcops, Jankuza, Tomballguy, Asphious, Chaheel Riens, FirecrackerDemon, BaronVonYiffington, Black Squirrel 2, Martin IIIa, Racerr, Lorson, RjwilmsiBot, Dewritech, GoingBatty, BrokenAnchorBot, TheTimesAreAChanging, ClueBot NG, Horseman16, Nestor1010, Yowanoreo, Hansen Sebastian, Dissident93, Bgibbs2, Cainamarques, Monkbot, MarioSonicU, Cawoodjosh, WikiSyn, BD2412bot and Anonymous: 56

- **The House of the Dead (video game)** *Source:* https://en.wikipedia.org/wiki/The_House_of_the_Dead_(video_game)?oldid=679078132 *Contributors:* Frecklefoot, SD6-Agent, Phil Boswell, LGagnon, MSGJ, Herbee, Mboverload, Ian Pugh, Fangz, Sam Hocevar, Spottedowl, Luigi, Zoganes, RossPatterson, Dmeranda, Zenohockey, Lampbane, DaveGorman, CyberSkull, Jtalledo, SeanDuggan, Alpha5099, Tony Sidaway, CherryMay, Hbdragon88, Combination, Joe Decker, N-Man, Purple Rose, Iainelder, YurikBot, Tznkai, RussBot, DT28, Splash, Randall Brackett, Mikeblas, Bobquest3, Empty2005, Kendricken, N. Harmonik, Closedmouth, SmackBot, DuoDeathscyther 02, Geoff B, Vega Vaikyuko, MisterHand, Benten, ThreeAnswers, GVnayR, Harryboyles, Evrain, JHunterJ, Crh66, Ace of Sevens, JoeBot, Angeldeb82, Mika1h, Jeremy Silver, Jesse Viviano, Rcldragon, Zamamee, Cydebot, Odie5533, Darker Samus, BetacommandBot, Thijs!bot, Sestren NK, Luigifan, Greg233, Dibol, X201, ThomasO1989, Xeno, Janadore, Bmbcali, Poitero, I love the film saw, Cube b3, CommonsDelinker, McDoobAU93, A Nobody, Notreallydavid, AntiSpamBot, SuperSonicTH, Gurzyb, Varnent, Guitaro193, WOSlinker, Lots42, Theraven329, ITurtle, Mr.NorCal55, Austriacus, Kael555, AMbot, Ongyefeng, TX55, Sword&Scythe, TaerkastUA, Beemer69, SubtleCynicism, Gloss, ClueBot, DarkFireYoshi, EoGuy, Mezigue, 718 Bot, Someone another, PixelBot, Iohannes Animosus, Mikaey, XLinkBot, Rankiri, Anticipation of a New Lover's Arrival, The, The Cre8r, Addbot, Megata Sanshiro, A. Falcao, Jjhendricks, BecauseWhy?, Lightbot, Luckas-bot, Yobot, Aqula125, KaedeCaleb, Obersachsebot, Donkeykong7, Capricorn42, CoolingGibbon, Chaheel Riens, Deltasim, FrescoBot, Sis and bro, Wallking, White Shadows, JCGDIMAIWAT, Lorson, Davidnagash~enwiki, Bossanoven, EmausBot, Starcheerspeaksnewslostwars, Irapekitties, Lacon432, NewRecruit, Koldcuts, Paulsbuck, TheTimesAreAChanging, ClueBot NG, Gemini J-Man, Easy4me, DanWiki2011, MerlIwBot, Strike Eagle, Aquario, BG19bot, Compfreak7, Bestboy7850, ChrisGualtieri, Just another guy in a suit, Dissident93, Cerabot~enwiki, Sotosbros, Mostofa-morjina, Eminem2001, Camyoung54, Rabbitgentleman, Heri Yeo, Zeddman123, Lord Monboddo, JaconaFrere, Kevieman94, AJFU and Anonymous: 280

- **Panzer Dragoon Saga** *Source:* https://en.wikipedia.org/wiki/Panzer_Dragoon_Saga?oldid=673497235 *Contributors:* Timwi, David Gerard, Mike Rosoft, Rich Farmbrough, Shadow Hog, Sobolewski, Bobrayner, Pypex, Nihiltres, Stormwatch, Hibana, ShadowHntr, Check two you, Zig973, NateDan, Druff, SmackBot, Cool3, Doktor Wilhelm, Chris the speller, Tghe-retford, NickSCFC, SubSeven, Sabrewing, CmdrObot, Zarex, Mika1h, Cydebot, James Duffy, Paddles, Uriptical, BetacommandBot, Thijs!bot, Voracious reader, X201, TarkusAB, Alphachimpbot, Appraiser, Janadore, Jimkemon, Xenoranger, Radicalfaith360, Frogacuda, Tomsega, Rei-bot, Niggahater, Connell66, Dawn Bard, JohnnyMrNinja, TypoBot, ImageRemovalBot, Agent Sleeper, Iuhkjhk87y678, Trivialist, Someone another, Thehelpfulone, Ost316, Addbot, Jjhendricks, Legobot, AnomieBOT, Materialscientist, LilHelpa, Erik9bot, FrescoBot, Ryo Suzuki, Corni Mueller, Martin IIIa, ZéroBot, SporkBot, EdoBot, TheTimesAreAChanging, Shaddim, Popcornduff, BG19bot, The Banner Turbo, Orenp, StealthMantis, Samwalton9, MadGuy7023, Dissident93, Sriharsh1234, Richolmes14, Kevieman94, Meppi64, Tripple-ddd, WikiSyn and Anonymous: 98

- **Dragon Force** *Source:* https://en.wikipedia.org/wiki/Dragon_Force?oldid=674199420 *Contributors:* Zundark, Chuq, K1Bond007, Aleron235, Nifboy, Kbh3rd, Ylee, Thunderbrand, Alansohn, CyberSkull, Ricky81682, Ericl234, Mandarax, FlaBot, TheDJ, Hibana, Dstln, EditingMachine, Rsrikanth05, Wiki alf, N. Harmonik, Closedmouth, SmackBot, Doktor Wilhelm, Sazabirules, SlimJim, Dross82, Heimstern, TJ Spyke, SubSeven, ShaleZero, BranStark, SlyDante, Wafulz, Lazulilasher, DakaSha, Cydebot, BetacommandBot, Epbr123, Marek69, Dayn, X201, Salavat, Luna Santin, Alphachimpbot, Dan D. Ric, Coreydragon, VoABot II, Patstuart, MartinBot, Roastytoast, Haxorelete, SharkD, NiGHTS into Dreams..., Radicalfaith360, Mr Wesker, Beem2, Hqb, Anonymous Dissident, Seb az86556, Xavcam, JT Dutch, Eloc Jcg, Gunmetal Angel, Kabapu, ClueBot, The Thing That Should Not Be, Trivialist, Lartoven, SoxBot, Aitias, 7, Anarkangel, Swindbot, GuardianGenko, Dauthus, Vandalizethis, DumZiBoT, InternetMeme, PseudoOne, Addbot, Willking1979, CanadianLinuxUser, LAAFan, Nyrock, Sardur, Lightbot, Yobot, MikeStuff, Mr T (Based), Orange highlighter, Frankenpuppy, Ragamath, Eugene-elgato, FrescoBot, Sexybait, Martin IIIa, Lacertilia the Magnificent, John of Reading, Evanh2008, Hydao, Msdudczak, TheTimesAreAChanging, NJ84, AvocatoBot, GKyTW, BattyBot, SNAAAAKE!!, Dexbot, Spreadsheeticus, Richolmes14, Kevieman94, LaytonPuzzle27 and Anonymous: 96

- **PowerSlave** *Source:* https://en.wikipedia.org/wiki/PowerSlave?oldid=679067539 *Contributors:* Rholton, Ciciban, Mboverload, DaveWebster, Notinasnaid, BuzzBomber, Shadow Hog, Pikawil, CyberSkull, Jtalledo, Drat, Dopefish, Kelly Martin, LOL, Sega381, Marudubshinki, BD2412, DiamondDave~enwiki, Bgwhite, PatCheng, N. Harmonik, DuoDeathscyther 02, Reedy, Mr. Pointy, Darklock, Cabe6403, Chris the speller, GVnayR, Marcus Brute, Ravimakkar, Khazar, Wickethewok, SubSeven, Mika1h, Zamamee, Cydebot, EssentialParadox, BetacommandBot, Lanky, X201, Dawkeye, Kariteh, Fennessy, Wdflake, Giantsandwichking, Plasticup, Eionm, Mr Wesker, Signalhead, WOSlinker, Comrade Graham, Austriacus, ImageRemovalBot, Martarius, Cobra 3000, Srirangav, Niceguyedc, Wiki libs, Project FMF, Rockk3r, Eik Corell, Ost316, Addbot, RandySavageFTW, LatitudeBot, Elbryan42, Numbo3-bot, LarryJeff, Lightbot, Tavatar, FrescoBot, HRoestBot, Jaguar, Orenburg1, Martin IIIa, Lorson, TheXenomorph1, ZéroBot, Asdfsfs, Clonehunter1, Shaddim, KLBot2, BG19bot, DmitryN, NukeofEarl, Hmainsbot1, Zachverb and Anonymous: 48

- **World Series Baseball (video game)** *Source:* https://en.wikipedia.org/wiki/World_Series_Baseball_(video_game)?oldid=677077330 *Contributors:* TMC1982, Bobo192, BD2412, Eubot, Coll7, Zagalejo, SMcCandlish, SmackBot, 32X, Bkrudy, GVnayR, Phoenixrod, Mika1h, Cydebot, PV250X, BetacommandBot, X201, Salavat, Caleson, RobotG, KidIncredible, RobJ1981, Waacstats, Rettetast, R'n'B, C. Foultz, ClueBot, Trivialist, MIjy~enwiki, Addbot, Materialscientist, Martin IIIa, Spilia4, Videogamefanatic5471 and Anonymous: 8

- **Sega Worldwide Soccer** *Source:* https://en.wikipedia.org/wiki/Sega_Worldwide_Soccer?oldid=675419097 *Contributors:* Mike Rosoft, Ynhockey, J 1982, TJ Spyke, Cydebot, Salavat, Hirolovesswords, Addbot, Fiftyquid, Luckas-bot, AnomieBOT, DeNoel, Martin IIIa, EmausBot, Paulsbuck, Dissident93, Hmainsbot1, BD2412bot and Anonymous: 4

- **Sonic X-treme** *Source:* https://en.wikipedia.org/wiki/Sonic_X-treme?oldid=679333383 *Contributors:* Norm, Tregoweth, WhisperToMe, Furrykef, Alan Liefting, Mboverload, Bobblewik, CALR, Rich Farmbrough, Xezbeth, Shadow Hog, Steerpike, Evice, Thunderbrand, CyberSkull, Jtalledo, Drat, Nintendo Maximus, ThomasWinwood, New Age Retro Hippie, Kelly Martin, Smoke, Starblind, Tabletop, Ppk01, ThomasHarte, Mandarax, Rjwilmsi, Aurochs, Koavf, SMC, Nick R, Dionyseus, FlaBot, Ian Pitchford, Stormwatch, Igordebraga, Bgwhite, FlareNUKE, Denjo, Epolk, Admiral Roo, RattleMan, Falcon9x5, VederJuda, N. Harmonik, Richardcavell, Conan-san, Theda, El cid the hero, YesIAmAnIdiot, BlazeHedgehog, H Hog, Jutl, Dr. R.K.Z, SmackBot, Cyberdude93, Khfan93, Reedy, JimmyBlackwing, Doktor Wilhelm, SynergyBlades, Tghe-retford, Onesimos, Zachkudrna18@yahoo.com, Jennytablina, GVnayR, Vusys, Phoenixj91, Derek R Bullamore, Seraphcrono, Teancum, IronGargoyle, TPIRFanSteve, JoeBot, Muéro, DKqwerty, Tifego, CmdrObot, Cydebot, ST47, Tawkerbot4, UnDeRsCoRe, BetacommandBot, Thijs!bot, 244pupil6, Lanky, Evan1109, X201, IceSage, Salavat, Sailor Angel, ChristinaCoffin, Quietpeoplerock, RobJ1981, Darklilac, Dragon DASH, Alphachimpbot, 77night77, Tactless, Bongwarrior, Hasek is the best, Wickedflea, Coffee4binky, Saberclaw, Brittany Ka, MartinBot, McDoobAU93, Little Professor, Noutakun, AntiSpamBot, Shadow Android, Juliancolton, Mr Wesker, 123lkik, RingtailedFox, TXiKiBoT, Pianoman13, Collision Cat, IndianCheese, Mlpfoster, Malcolmxl5, Chromaticity, Leviathan2000, Staticz, JohnnyMrNinja, Altzinn, SuperSonic56, ImageRemovalBot, ClueBot, Elephant Talk, Red Phoenix, Mild Bill Hiccup, ChaosAngelZero, Bokan, Razorflame, Randomran, HeaveTheClay, Gyozilla, Tezero, Deespence2929, Polo83, MystBot, Kingplatypus, Addbot, Megata Sanshiro, Mac Dreamstate, J.S.thehedgehog, MuZemike, The muramasa, Legobot, Yobot, AnomieBOT, JackieBot, Piano non troppo, Tvfan01nine, Darkninja98, Citation bot, RockJuno, TheDumbening, Nasnema, Sergecross73, FirecrackerDemon, Untick, PickingGold12, Agbwiki, BigDwiki, Secret Saturdays, Lightlowemon, Vrenator, ZZanimar, DARTH SIDIOUS 2, John of Reading, Radix Z, OhstaWiki, TheTimesAreAChanging, Ednoone, Satellizer, Despatche, Fig2002, Helpful Pixie Bot, Bigfatpiggy, Josue58423, Zeldajiggmin, Gabriel Yuji, Mslurr, LoneWolf1992, Cetange23, Dobie80, Gschandler, SegaKing247, Medachod, Кирилл Ерин, Sonic N800, NGreen10774, Maiathebest7878787878, Qunera and Anonymous: 212

- **List of Sega Saturn games** *Source:* https://en.wikipedia.org/wiki/List_of_Sega_Saturn_games?oldid=680850609 *Contributors:* Kingturtle, Shoecream, Fredrik, LGagnon, Peter L, Nifboy, Misterkillboy, Tubedogg, Abdull, D6, DaveWebster, Shadow Hog, ZeroOne, Steerpike, MIT Trekkie, Marasmusine, Woohookitty, LOL, Combination, BD2412, MrLeo, KramarDanIkabu, Noirish, Rjwilmsi, Jelly Soup, Kakonator, Stormwatch, Joerger, Dstln, YurikBot, Royalbroil, Buuneko, Welsh, Norbi, JJBunks, 2fort5r, DCEvoCE, Noidner, Merkuri, Jcbarr, ZS, Jimmy-Blackwing, Sarujo, Cabe6403, Tghe-retford, Kermit blue, Bazonka, Chlewbot, GVnayR, Ifrit, Jeffersoncpost, Atirage, E-Kartoffel, Eliashc, TJ Spyke, JeffW, Angeldeb82, ChrisCork, Cyrus XIII, Rbl.134, Jac16888, Cydebot, Khatru2, Odie5533, Rhe br, Chachilongbow, Thijs!bot, Truten, TangentCube, Salavat, Fru1tbat, Wamu, JAnDbot, Dm82, MegX, Angel,Isaac, Magioladitis, TVfanatic2K, JamesBWatson, Thibbs, Nikdog, R'n'B, Atama, Mairebleu, Mr Wesker, Grahammer, Sonicsean89, WOSlinker, Maxtremus, January2007, Richj1983, PhoenixML, Legoktm, 31Gabe, Seanpaulk, NJChristian07, Jesus.arnold, Grolim, Ildfisk~enwiki, Fratrep, JohnnyMrNinja, JL-Bot, Halo2, Sfan00 IMG, Plastikspork, EoGuy, Chessage, Red Phoenix, Aylesburyape, Niceguyedc, Canis Lupus, ImpactMegaton, Xeniczone, Ost316, Addbot, Yobot, Fraggle81, Diegowar, Mcjakeqcool, Nicholsonadam, Materialscientist, Grey ghost, Xqbot, The Banner, RibotBOT, DeNoel, FrescoBot, Jmdeleon, Redrose64, TomasHA-SK, Black Squirrel 2, Martin IIIa, Callanecc, TL565, MAXXX-309, WildBot, Koppapa, EmausBot, Ragowit, GoingBatty, TuHan-Bot, Jeffmendoza, Hydao, Overtheriver565, Jamesster.LEGO, Vittupaska123, TheTimesAreAChanging, Horseman16, MelbourneStar, Baldy Bill, Vonicemen69, BG19bot, Pdiddyjr, SuperSherbet, DPL bot, The1337gamer, Jbdodson, SprinterBot, Dexbot, Dissident93, Incrediblehark, BDE1982, Palavar, VideoGameUpdaterUK, Fushindiz, Kevieman94, Katana99, Dmatteng, Mr.gangsta, Kevbrook, MarioSonicU, Tokeepongaming, Supdiop and Anonymous: 153

- **History of video game consoles (fifth generation)** *Source:* https://en.wikipedia.org/wiki/History_of_video_game_consoles_(fifth_generation) ?oldid=678725410 *Contributors:* Damian Yerrick, Shii, Mrwojo, Edward, DopefishJustin, Tompagenet, KAMiKAZOW, Kimiko, Andrevan, Tedius Zanarukando, Greenrd, Xiaodai~enwiki, K1Bond007, Ed g2s, Morn, Wiwaxia, Fredrik, Lowellian, Auric, Benc, Boarder8925, Alexwcovington, Jacoplane, Gtrmp, Ciciban, Andrew Weintraub, Guanaco, Masken, Mboverload, Gracefool, Unknownwarrior33, Neilc, Thewikipedian, Chowbok, Josquius, D3v4st4t0r, Tubedogg, Generica, Scottk, Erc, Vague Rant, Pixel8, LindsayH, Shadow Hog, Indrian, Bender235, Sockatume, CanisRufus, RoyBoy, Kctsang, ERW1980, Cmdrjameson, VoiceOfReason, Pikawil, Greyhead, Diceman, Tyan23, Rje, Jason One, Jtalledo, InShaneee, Seancdaug, DavidCWG, Gbeeker, TheDotGamer, Itschris, Drat, MIT Trekkie, Kitch, Red dwarf, Chardish, Kelly Martin, Firsfron, Alvis, Woohookitty, Blackeagle, LOL, TomTheHand, Misterjta, Madchester, ThomasHarte, Combination, Marvelvsdc, Holek, Seishirou Sakurazuka, Captain Spyro, BD2412, David Levy, RadioActive~enwiki, Rjwilmsi, Nick R, Vegaswikian, ABot, Remurmur, Rangek, Dionyseus, SchuminWeb, JeremyMcCracken, ShotokanTuning, BcRIPster, Y2j420, ApolloBoy, Benlisquare, Igordebraga, Sonic Mew, Measure, Wavelength, Hairy Dude, Kafziel, XKuei, Muchness, Allister MacLeod, Kevs, Aaron Walden, Wgungfu, Pelago, Royalbroil, Gunnar-Rene, Anomie, GSK, JHCaufield, Oakster, Gadget850, CLW, Strolls, Crisco 1492, PTSE, Cpl.Punishment, YolanCh, Th1rt3en, DisambigBot, Nall, Doom127, AMHR285, Aresmo, Tom Morris, Ryūkotsusei, SmackBot, Merkuri, Renegadeviking, Dieboybun, Jagged 85, Kittynboi, ZS, Mauls, Betacommand, Poulsen, Slo-mo, J.L.Main, GoldDragon, 32X, Kfroog, Mdwh, Apple2gs, George Ho, Killerisation, MisterHand, OrphanBot, Lostinlodos, Nayl, Nyletak, Irish Souffle, Regulus Tera, Super box A, Komojo, AMaltese, Mini-Geek, AlexJ, Tehw1k1, Evlekis, BlackTerror, Mcdoomer~enwiki, Anss123, Axem Titanium, AlphaTwo, Arnoox, Takuthehedgehog, 041744, Berenlazarus, Gormanly, SQGibbon, Thatcher, Eliashc, TJ Spyke, SubSeven, Ace of Sevens, Twas Now, Octane, ReYo, Phoenixrod, FairuseBot, Tawkerbot2, DKqwerty, Evangelion883, CmdrObot, True 505, Cyrus XIII, Nczempin, Mika1h, Raptor007, Neelix, Ravensfan5252, Auger Martel, Moonbat, Cydebot, Tamagon, MindWraith, PV250X, Dancter, Neoforma, Chrislk02, DiScOrD tHe LuNaTiC, GameJunkieJim, Lord Hawk, 803290, Dvferret, Silver Edge, Biggman15, Sion8, Monkeyangst, Jhsounds, Spencer, Myanw, PresN, Porlob, Coreydragon, Gavia immer, Moviemaniacx, Satoru003, Y2kcrazyjoker4, TVfanatic2K, Nyttend, Wesker85, Robotman1974, Shade 5154, Osh33m, Wdflake, Willy105, Hemidemisemiquaver, SpecialWindler, Hdt83, MartinBot, CommonsDelinker, EdBever, J.delanoy, Trusilver, JJLESDUDE, Adavidb, Richiekim, SharkD, Plasticup, ShadowLaguna, STBotD, C. Foultz, Tygrrr, Mincebert, P924s88, S, Useight, VolkovBot, NShadow1993, Vgexpert, Masaruemoto, RingtailedFox, HeihachiMishima, NYkid0709, Wiimonk, JayC, Cammelspit, FourteenDays, OfficerPhil, Jonnyf88, Djmckee1, AlleborgoBot, 31Gabe, SieBot, Brenont, Winchelsea, Nebulousecho, Merotoker1, Zurqoxn, Spock2266, Wrathoftheafe, StaticGull, DeepQuasar, SpinyMcSpleen, Denisarona, Willy, your mate, ImageRemovalBot, The amazing Magic idiot, Martarius, Sfan00 IMG, ClueBot, Canes5tk, Cobra 3000, Badger Drink, The Thing That Should Not Be, Madskunk, Red Phoenix, Eab969, HweeWieng, JTBX, Niceguyedc, Thekoyaanisqatsi, Bannedboy, Bobby Tables, Muhandes, Arjayay, Iohannes Animosus, Holothurion, IvanTheRussian, Thingg, Project FMF, InternetMeme, XLinkBot, Wolfer68, Ost316, KyleFN, WikHead, Nuxius, Addbot, WishfireOmega, Nintendog master 54, Download, Pug007, Krano, Arbitrarily0, JohnnyPolo24, Yobot, Incog88, KamikazeBot, Ayrton Prost, Nyat, AnomieBOT, TJD2, Jim1138, M3549S3X7AL, PhaseChanger, Mcjakeqcool, Danno uk, Xqbot, Warmpuppy2, Tomflaherty, Mlwgsgis1487, Mario777Zelda, 1779Days, Patrykwiki, PigFlu Oink, Typingwestern015, Rapsar, Calmer Waters, BigDwiki, DMJC, RedBot, Luph25, Anarchy's eye, Secret Saturdays, Full-date unlinking bot, Politics-man, Cnwilliams, Mariacer Cervantes, Martin IIIa, Wo.luren, Tbhotch, Reach Out to the Truth, Found A Dojo, EmausBot, WikitanvirBot, Niwi3, G&CP, Somebody500, Iamcool234, Illegitimate Barrister, Liquidmetalrob, Josve05a, Sengokucannon, Pank x, Δ, Asdfsfs, Mentibot, Evan-Amos, TheTimesAreAChanging, ClueBot NG, Horseman16, Snotbot, Widr, Helpful Pixie Bot, BG19bot, Mohamed CJ, NukeofEarl, Queen of Awesome, Supernerd11, Jeancey, Harizotoh9, Matt Mohandas Gandhi, GSJones1994, Drecoll, Golfballz, Dobie80, Dexbot, Mogism, Guiletheme, Reatlas, PurpleMesa, SegaKing247, Neilandio, MuddyMaestro, Billybob2002, Comp.arch, Gixce93, TheWildOnesEu, Oranjelo100, 636Buster, SpeedBlur, Sheepkicks, TNOA, Noctis1436, Monkbot, EEVVYY, Leeroyhim, KombatPolice, Retroking1981, Danny S. Domokos, BustaBunny, MonkeyWithGlasses44, DJV11181988, StefanAshtonFrank, GodOFGamers72, Cartakes, TomatoHentai, BD2412bot and Anonymous: 474

- **List of Sega arcade system boards** *Source:* https://en.wikipedia.org/wiki/List_of_Sega_arcade_system_boards?oldid=678450495 *Contributors:* KAMiKAZOW, ZoeB, Furrykef, Saltine, Joy, TriMesh, Ciciban, Mamizou, Bumm13, Tooki, DmitryKo, Simon Fenney, YUL89YYZ, Indrian, Bender235, Mithent, Woohookitty, Tabletop, Rjwilmsi, Nick R, Jonny2x4, Lavenderbunny, SmackBot, Pellucidity, Jagged 85, Darklock, Parrothead1983, Chris the speller, Thumperward, OrangeDog, Frap, Jacob Poon, OrphanBot, Jhonsrid, Martijn Hoekstra, WhosAsking, Anss123, Green Giant, Tlesher, Phuzion, TJ Spyke, Gaunt, CmdrObot, Mika1h, Jac16888, Odie5533, Akadewboy, Electron9, X201, Mentifisto, Markthemac, Jklsemicolon, Robivy64, Yaca2671~enwiki, Nono64, Tntdj, SpigotMap, DeeKay64, Signalhead, Nomaxxx, Espiox, Nagy, Josh7289, Wageslave, Lightmouse, PbBot, Fratrep, Svick, Cyfal, ImageRemovalBot, Martarius, Sonictrey, Hippo99, Rilak, Czarkoff, Red Phoenix, Niceguyedc, Trivialist, 718 Bot, PatLTornado, Usx9, Sun Creator, Gunnar.offel, InternetMeme, Bridies, XLinkBot, Ost316, WikHead, Debresser, Lightbot, Smeagol 17, Shinobi MVP, Aruto k, X-Pilot, Yobot, Tohd8BohaithuGh1, KaosuKaiari, Dandy Sephy, Daniel7066, AnomieBOT, WorkingBeaver, Segaamusements, Piano non troppo, GanicoGSx, Elm-39, NeoDoubleGames, Junkcops, J04n, FrescoBot, Ibwonton, Onel5969, RjwilmsiBot, Whanger.choi, Mattboston, Becritical, Dewritech, Δ, L Kensington, Arosio Stefano, TheTimesAreAChanging, ClueBot NG, Horseman16, Matthiaspaul, Korrawit, Faster2010, BG19bot, Mahcann, BattyBot, Dissident93, SoledadKabocha, DarkToonLink, Comp.arch, Makusensu, Monkbot, Tripple-ddd and Anonymous: 169

## 5.2 Images

- **File:315-5687_01.jpg** *Source:* https://upload.wikimedia.org/wikipedia/commons/f/f9/315-5687_01.jpg *License:* CC-BY-SA-3.0 *Contributors:* ? *Original artist:* ?

- **File:315-5689_01.jpg** *Source:* https://upload.wikimedia.org/wikipedia/commons/1/1e/315-5689_01.jpg *License:* CC-BY-SA-3.0 *Contributors:* ? *Original artist:* ?

- **File:315-5690_VDP2_01.jpg** *Source:* https://upload.wikimedia.org/wikipedia/commons/4/4e/315-5690_VDP2_01.jpg *License:* CC-BY-SA-3.0 *Contributors:* ? *Original artist:* ?

- **File:3DO-FZ1-Console-Set.jpg** *Source:* https://upload.wikimedia.org/wikipedia/commons/5/56/3DO-FZ1-Console-Set.jpg *License:* CC BY-SA 3.0 *Contributors:* Own work *Original artist:* Evan-Amos

- **File:4._Magician_Screen_(Old).PNG** *Source:* https://upload.wikimedia.org/wikipedia/en/7/7e/4._Magician_Screen_%28Old%29.PNG *License:* ? *Contributors:*
Screenshot presumably taken by uploader.
*Original artist:* ?

- **File:Atari-Jaguar-Console-Set.jpg** *Source:* https://upload.wikimedia.org/wikipedia/commons/f/fc/Atari-Jaguar-Console-Set.jpg *License:* CC BY-SA 3.0 *Contributors:* Own work *Original artist:* Evan-Amos

- **File:Baseball.svg** *Source:* https://upload.wikimedia.org/wikipedia/commons/9/92/Baseball.svg *License:* CC0 *Contributors:* http://openclipart.org/ *Original artist:* vedub4us

- **File:Burning_force_pcb.PNG** *Source:* https://upload.wikimedia.org/wikipedia/commons/c/ca/Burning_force_pcb.PNG *License:* Public domain *Contributors:* Transferred from en.wikipedia; transferred to Commons by User:Liftarn using CommonsHelper.
*Original artist:* Robivy64 (talk). Original uploader was Robivy64 at en.wikipedia

- **File:CD-Rom-Drive'{}s_Laser.jpg** *Source:* https://upload.wikimedia.org/wikipedia/commons/5/5e/CD-Rom-Drive%27s_Laser.jpg *License:* CC BY-SA 3.0 *Contributors:* eigenes Bild
own picture *Original artist:* Matthias Zepper

- **File:CD_laser_assembly1.jpg** *Source:* https://upload.wikimedia.org/wikipedia/commons/f/f0/CD_laser_assembly1.jpg *License:* CC-BY-SA-3.0 *Contributors:* ? *Original artist:* Flip619 at English Wikipedia

- **File:Cd-rom-drive-reading-head-movement.gif** *Source:* https://upload.wikimedia.org/wikipedia/commons/7/7e/Cd-rom-drive-reading-head-movement.gif *License:* Public domain *Contributors:* eigene Animation
own work *Original artist:* Matthias Zepper

- **File:Chaos_emeralds.svg** *Source:* https://upload.wikimedia.org/wikipedia/commons/f/f0/Chaos_emeralds.svg *License:* CC BY-SA 3.0 *Contributors:* Own work *Original artist:* Gringer

- **File:Commons-logo.svg** *Source:* https://upload.wikimedia.org/wikipedia/en/4/4a/Commons-logo.svg *License:* ? *Contributors:* ? *Original artist:* ?

- **File:Denso-SH2.jpg** *Source:* https://upload.wikimedia.org/wikipedia/commons/3/38/Denso-SH2.jpg *License:* Public domain *Contributors:* Sony DSC-H9 Digital Camera *Original artist:* GMPX

- **File:Dragon_Force_Coverart.png** *Source:* https://upload.wikimedia.org/wikipedia/en/b/b0/Dragon_Force_Coverart.png *License:* Fair use *Contributors:*
  http://www.gamefaqs.com/saturn/197149-dragon-force/images/box-42036 *Original artist:* ?
- **File:Edit-clear.svg** *Source:* https://upload.wikimedia.org/wikipedia/en/f/f2/Edit-clear.svg *License:* Public domain *Contributors:* The *Tango! Desktop Project. Original artist:*
  The people from the Tango! project. And according to the meta-data in the file, specifically: "Andreas Nilsson, and Jakub Steiner (although minimally)."
- **File:Factory_1b.svg** *Source:* https://upload.wikimedia.org/wikipedia/commons/b/b6/Factory_1b.svg *License:* CC-BY-SA-3.0 *Contributors:* PNG version on the English Wikipedia *Original artist:* Dtbohrer, updated to SVG by Tomtheman5
- **File:Flag_of_Europe.svg** *Source:* https://upload.wikimedia.org/wikipedia/commons/b/b7/Flag_of_Europe.svg *License:* Public domain *Contributors:*
- File based on the specification given at [1]. *Original artist:* User:Verdy p, User:-xfi-, User:Paddu, User:Nightstallion, User:Funakoshi, User:Jeltz, User:Dbenbenn, User:Zscout370
- **File:Flag_of_Japan.svg** *Source:* https://upload.wikimedia.org/wikipedia/en/9/9e/Flag_of_Japan.svg *License:* PD *Contributors:* ? *Original artist:* ?
- **File:Flag_of_the_United_States.svg** *Source:* https://upload.wikimedia.org/wikipedia/en/a/a4/Flag_of_the_United_States.svg *License:* PD *Contributors:* ? *Original artist:* ?
- **File:Folder_Hexagonal_Icon.svg** *Source:* https://upload.wikimedia.org/wikipedia/en/4/48/Folder_Hexagonal_Icon.svg *License:* Cc-by-sa-3.0 *Contributors:* ? *Original artist:* ?
- **File:Gamepad.svg** *Source:* https://upload.wikimedia.org/wikipedia/en/b/be/Gamepad.svg *License:* ? *Contributors:* ? *Original artist:* ?
- **File:Genesis_controller.png** *Source:* https://upload.wikimedia.org/wikipedia/commons/0/0a/Genesis_controller.png *License:* Public domain *Contributors:* Own work *Original artist:* Joel Trottier
- **File:HD6417095_01.jpg** *Source:* https://upload.wikimedia.org/wikipedia/commons/4/4f/HD6417095_01.jpg *License:* CC-BY-SA-3.0 *Contributors:* Own work *Original artist:* Yaca2671 撮影
- **File:Hitachi_SH3.jpg** *Source:* https://upload.wikimedia.org/wikipedia/commons/e/ed/Hitachi_SH3.jpg *License:* Public domain *Contributors:* Transferred from en.wikipedia; transferred to Commons by User:Sreejithk2000 using CommonsHelper. *Original artist:* Original uploader was Saxbryn at en.wikipedia
- **File:KL_Hitachi_HD68000.jpg** *Source:* https://upload.wikimedia.org/wikipedia/commons/a/ac/KL_Hitachi_HD68000.jpg *License:* GFDL *Contributors:* CPU collection Konstantin Lanzet *Original artist:* Konstantin Lanzet
- **File:KL_Motorola_68EC000_PLCC.jpg** *Source:* https://upload.wikimedia.org/wikipedia/commons/a/a5/KL_Motorola_68EC000_PLCC.jpg *License:* CC BY 3.0 *Contributors:* CPU collection *Original artist:* Konstantin Lanzet
- **File:KL_Motorola_MC68000_CLCC.jpg** *Source:* https://upload.wikimedia.org/wikipedia/commons/2/2e/KL_Motorola_MC68000_CLCC.jpg *License:* CC BY 3.0 *Contributors:* CPU collection *Original artist:* Konstantin Lanzet
- **File:KL_Motorola_MC68000_PLCC.jpg** *Source:* https://upload.wikimedia.org/wikipedia/commons/4/48/KL_Motorola_MC68000_PLCC.jpg *License:* CC BY 3.0 *Contributors:* CPU collection *Original artist:* Konstantin Lanzet
- **File:KL_Thomson_TS68000.jpg** *Source:* https://upload.wikimedia.org/wikipedia/commons/0/08/KL_Thomson_TS68000.jpg *License:* GFDL *Contributors:* CPU collection Konstantin Lanzet *Original artist:* Konstantin Lanzet
- **File:Motorola_68000_die.JPG** *Source:* https://upload.wikimedia.org/wikipedia/commons/f/fe/Motorola_68000_die.JPG *License:* CC BY 3.0 *Contributors:* Own work *Original artist:* Pauli Rautakorpi
- **File:N64-Console-Set.jpg** *Source:* https://upload.wikimedia.org/wikipedia/commons/1/11/N64-Console-Set.jpg *License:* Public domain *Contributors:* Own work *Original artist:* Evan-Amos
- **File:NiGHTs_into_Dreams,_Saturn_version,_Spring_Valley.jpg** *Source:* https://upload.wikimedia.org/wikipedia/en/e/e3/NiGHTs_into_Dreams%2C_Saturn_version%2C_Spring_Valley.jpg *License:* Fair use *Contributors:* http://retrowaretv.com/still-loading-games-to-travel-with/ *Original artist:* Sonic Team/Sega
- **File:NintendoStack.jpg** *Source:* https://upload.wikimedia.org/wikipedia/commons/3/3a/NintendoStack.jpg *License:* CC-BY-SA-3.0 *Contributors:* Transferred from en.wikipedia to Commons. *Original artist:* Wuffyz at English Wikipedia Later versions were uploaded by Shawnc at en.wikipedia.
- **File:PSX-Console-wController.jpg** *Source:* https://upload.wikimedia.org/wikipedia/commons/3/39/PSX-Console-wController.jpg *License:* Public domain *Contributors:* Own work *Original artist:* Evan-Amos
- **File:PanzerDragoonSagaBox.jpg** *Source:* https://upload.wikimedia.org/wikipedia/en/6/64/PanzerDragoonSagaBox.jpg *License:* Fair use *Contributors:*
  SEGA Corporation
  *Original artist:* ?
- **File:Pds_screen.jpg** *Source:* https://upload.wikimedia.org/wikipedia/en/c/c0/Pds_screen.jpg *License:* Fair use *Contributors:*
  Images from the game can be obtained from Sega. *Original artist:* ?
- **File:Portal-puzzle.svg** *Source:* https://upload.wikimedia.org/wikipedia/en/f/fd/Portal-puzzle.svg *License:* Public domain *Contributors:* ? *Original artist:* ?

- **File:Print_vs._bytes.jpg** *Source:* https://upload.wikimedia.org/wikipedia/commons/f/f0/Print_vs._bytes.jpg *License:* Public domain *Contributors:* Own work *Original artist:* Necessary Evil

- **File:Question_book-new.svg** *Source:* https://upload.wikimedia.org/wikipedia/en/9/99/Question_book-new.svg *License:* Cc-by-sa-3.0 *Contributors:*

  Created from scratch in Adobe Illustrator. Based on Image:Question book.png created by User:Equazcion *Original artist:*

  Tkgd2007

- **File:SEGA_logo.svg** *Source:* https://upload.wikimedia.org/wikipedia/commons/1/13/SEGA_logo.svg *License:* Public domain *Contributors:* Transferred from en.wikipedia to Commons by TFCforever. *Original artist:* The original uploader was Tkgd2007 at English Wikipedia

- **File:SH7091_01.jpg** *Source:* https://upload.wikimedia.org/wikipedia/commons/1/11/SH7091_01.jpg *License:* CC-BY-SA-3.0 *Contributors:* ? *Original artist:* ?

- **File:Sega-Saturn-3D-Controller.jpg** *Source:* https://upload.wikimedia.org/wikipedia/commons/b/b8/Sega-Saturn-3D-Controller.jpg *License:* Public domain *Contributors:* Own work *Original artist:* Evan-Amos

- **File:Sega-Saturn-Backup.jpg** *Source:* https://upload.wikimedia.org/wikipedia/commons/0/07/Sega-Saturn-Backup.jpg *License:* CC BY-SA 3.0 *Contributors:* Own work *Original artist:* Evan-Amos

- **File:Sega-Saturn-Console-Set-Mk1.jpg** *Source:* https://upload.wikimedia.org/wikipedia/commons/a/a8/Sega-Saturn-Console-Set-Mk1.jpg *License:* Public domain *Contributors:* Own work *Original artist:* Evan-Amos

- **File:Sega-Saturn-Console-Set-Mk1.png** *Source:* https://upload.wikimedia.org/wikipedia/commons/2/20/Sega-Saturn-Console-Set-Mk1.png *License:* Public domain *Contributors:* Own work *Original artist:* Evan-Amos

- **File:Sega-Saturn-Controller-Mk-I-NA-FL.jpg** *Source:* https://upload.wikimedia.org/wikipedia/commons/7/7c/Sega-Saturn-Controller-Mk-I-NA-FL.jpg *License:* Public domain *Contributors:* Own work *Original artist:* Evan-Amos

- **File:Sega-Saturn-Controller-NA-Mk-II-FL.jpg** *Source:* https://upload.wikimedia.org/wikipedia/commons/e/e0/Sega-Saturn-Controller-NA-Mk-II-FL.jpg *License:* Public domain *Contributors:* Own work *Original artist:* Evan-Amos

- **File:Sega-Saturn-JP-Mk1-Console-Set.jpg** *Source:* https://upload.wikimedia.org/wikipedia/commons/9/95/Sega-Saturn-JP-Mk1-Console-Set.jpg *License:* Public domain *Contributors:* Own work *Original artist:* Evan-Amos

- **File:Sega-Saturn-JP-Mk2-Console-Set.jpg** *Source:* https://upload.wikimedia.org/wikipedia/commons/c/c3/Sega-Saturn-JP-Mk2-Console-Set.jpg *License:* Public domain *Contributors:* Own work *Original artist:* Evan-Amos

- **File:Sega-Saturn-JP-Mk2-Console-Set.png** *Source:* https://upload.wikimedia.org/wikipedia/commons/2/2d/Sega-Saturn-JP-Mk2-Console-Set.png *License:* Public domain *Contributors:* Own work *Original artist:* Evan-Amos

- **File:Sega-Saturn-Motherboard.jpg** *Source:* https://upload.wikimedia.org/wikipedia/commons/9/9d/Sega-Saturn-Motherboard.jpg *License:* Public domain *Contributors:* Own work *Original artist:* Evan-Amos

- **File:Sega-Saturn-Multitap.jpg** *Source:* https://upload.wikimedia.org/wikipedia/commons/2/25/Sega-Saturn-Multitap.jpg *License:* Public domain *Contributors:* Own work *Original artist:* Evan-Amos

- **File:Sega-Saturn-NetLink.jpg** *Source:* https://upload.wikimedia.org/wikipedia/commons/2/2a/Sega-Saturn-NetLink.jpg *License:* Public domain *Contributors:* Own work *Original artist:* Evan-Amos

- **File:SegaSaturn.png** *Source:* https://upload.wikimedia.org/wikipedia/en/4/42/SegaSaturn.png *License:* Fair use *Contributors:*

  http://cache.kotaku.com/assets/resources/2007/01/SegaSaturn.gif *Original artist:* ?

- **File:SegaSaturnjp.png** *Source:* https://upload.wikimedia.org/wikipedia/en/f/f2/SegaSaturnjp.png *License:* Fair use *Contributors:*

  http://sega.jp/archive/segahard/ss *Original artist:* ?

- **File:Sega_Annual_Icome(Loss)_1993-2004.svg** *Source:* https://upload.wikimedia.org/wikipedia/commons/e/e3/Sega_Annual_Icome%28Loss%29_1993-2004.svg *License:* Public domain *Contributors:* Own work *Original artist:* Raffage

- **File:Sega_Rally_flyer.jpg** *Source:* https://upload.wikimedia.org/wikipedia/en/e/eb/Sega_Rally_flyer.jpg *License:* ? *Contributors:* ? *Original artist:* ?

- **File:Segastvpcb.jpg** *Source:* https://upload.wikimedia.org/wikipedia/commons/1/1e/Segastvpcb.jpg *License:* CC BY-SA 3.0 *Contributors:* Own work (Robivy64)

  *Original artist:* Robert Ivy (Robivy64 at en.wikipedia)

- **File:Slot_machines_at_Wookey_Hole_Caves.JPG** *Source:* https://upload.wikimedia.org/wikipedia/commons/3/34/Slot_machines_at_Wookey_Hole_Caves.JPG *License:* Public domain *Contributors:* Own work *Original artist:* Rodw

- **File:Sonic_1991.png** *Source:* https://upload.wikimedia.org/wikipedia/en/d/df/Sonic_1991.png *License:* Fair use *Contributors:*

  A scan.

  *Original artist:* ?

- **File:Sonic_X-treme_engine_test_screenshot.png** *Source:* https://upload.wikimedia.org/wikipedia/en/9/96/Sonic_X-treme_engine_test_screenshot.png *License:* Fair use *Contributors:*

  Self-made.

  *Original artist:* ?

- **File:Sony-Internal-PC-DVD-Drive-Opened.jpg** *Source:* https://upload.wikimedia.org/wikipedia/commons/f/f3/Sony-Internal-PC-DVD-Drive-Opened.jpg *License:* Public domain *Contributors:* Own work *Original artist:* Evan-Amos

- **File:Star_empty.svg** *Source:* https://upload.wikimedia.org/wikipedia/commons/4/49/Star_empty.svg *License:* CC BY-SA 2.5 *Contributors:* Made with Inkscape from Stars615.svg: &lt;a href='//commons.wikimedia.org/wiki/File:Stars615.svg' class='image'&gt;&lt;img alt='Stars615.svg' src='https://upload.wikimedia.org/wikipedia/commons/thumb/7/72/Stars615.svg/96px-Stars615.svg.png' width='96' height='17' srcset='https://upload.wikimedia.org/wikipedia/commons/thumb/7/72/Stars615.svg/144px-Stars615.svg.png 1.5x, https://upload.wikimedia.org/wikipedia/commons/thumb/7/72/Stars615.svg/192px-Stars615.svg.png 2x' data-file-width='640' data-file-height='110' /&gt;&lt;/a&gt;. *Original artist:* This vector image was created with Inkscape by Conti from the original images by RedHotHeat, and then manually edited.

- **File:Star_full.svg** *Source:* https://upload.wikimedia.org/wikipedia/commons/5/51/Star_full.svg *License:* Public domain *Contributors:* Made with Inkscape from Image:Stars615.svg. *Original artist:* User:Conti from the original images by User:RedHotHeat

- **File:Star_half.svg** *Source:* https://upload.wikimedia.org/wikipedia/commons/8/81/Star_half.svg *License:* CC BY-SA 2.5 *Contributors:* Made with Inkscape from Image:Stars615.svg. *Original artist:* User:Conti

- **File:Sxtreme-jadegully.jpg** *Source:* https://upload.wikimedia.org/wikipedia/en/2/29/Sxtreme-jadegully.jpg *License:* Fair use *Contributors:* Secrets of Sonic Team *Original artist:* ?

- **File:Symbol_book_class2.svg** *Source:* https://upload.wikimedia.org/wikipedia/commons/8/89/Symbol_book_class2.svg *License:* CC BY-SA 2.5 *Contributors:* Mad by Lokal_Profil by combining: *Original artist:* Lokal_Profil

- **File:Tsutenkaku2.jpg** *Source:* https://upload.wikimedia.org/wikipedia/commons/f/f8/Tsutenkaku2.jpg *License:* CC-BY-SA-3.0 *Contributors:* Momopy *Original artist:* Momopy

- **File:Video_game_history_icon.svg** *Source:* https://upload.wikimedia.org/wikipedia/commons/9/96/Video_game_history_icon.svg *License:* CC BY 2.5 *Contributors:* Compilation of File:Pillar ionic.svg & File:Gamepad.svg *Original artist:* User:JohnnyMrNinja based on work by w:David Vignoni & user:Helix84

- **File:Virtua-fighter-2-box.jpg** *Source:* https://upload.wikimedia.org/wikipedia/en/0/01/Virtua-fighter-2-box.jpg *License:* Fair use *Contributors:*

  http://flyers.arcade-museum.com/?page=flyer&db=videodb&id=1313&image=1 *Original artist:* ?

- **File:Volant_Sega_Saturn.jpg** *Source:* https://upload.wikimedia.org/wikipedia/commons/c/c8/Volant_Sega_Saturn.jpg *License:* CC BY-SA 3.0 *Contributors:* Own work *Original artist:* César3D

- **File:Wiki_letter_w_cropped.svg** *Source:* https://upload.wikimedia.org/wikipedia/commons/1/1c/Wiki_letter_w_cropped.svg *License:* CC-BY-SA-3.0 *Contributors:*

- Wiki_letter_w.svg *Original artist:* Wiki_letter_w.svg: Jarkko Piiroinen

- **File:Wiktionary-logo.svg** *Source:* https://upload.wikimedia.org/wikipedia/commons/e/ec/Wiktionary-logo.svg *License:* CC BY-SA 3.0 *Contributors:* ? *Original artist:* ?

- **File:XC68000.agr.jpg** *Source:* https://upload.wikimedia.org/wikipedia/commons/2/28/XC68000.agr.jpg *License:* CC-BY-SA-3.0 *Contributors:* ? *Original artist:* ?

## 5.3   Content license

- Creative Commons Attribution-Share Alike 3.0

www.ingramcontent.com/pod-product-compliance
Lightning Source LLC
Chambersburg PA
CBHW082302200526
45168CB00017B/2518